LEARNING GUIDE WITH INTEGRATED REVIEW WORKSHEETS

C. BRAD DAVIS

COLLEGE ALGEBRA WITH INTEGRATED REVIEW

Robert Blitzer

Miami Dade College

PEARSON

Boston Columbus Indianapolis New York San Francisco Upper Saddle River
Amsterdam Cape Town Dubai London Madrid Milan Munich Paris Montreal Toronto
Delhi Mexico City São Paulo Sydney Hong Kong Seoul Singapore Taipei Tokyo

Reproduced by Pearson from electronic files supplied by the author.

Publishing as Pearson, 75 Arlington Street, Boston, MA 02116.

ISBN-13: 978-0-321-97924-7
ISBN-10: 0-321-97924-9

1 2 3 4 5 6 CRK 17 16 15 14

www.pearsonhighered.com

PEARSON

Table of Contents

About this *Learning Guide*

Dear Student,

We are glad you have decided to read this introduction to your *Learning Guide*. By looking at the following great questions that your classmates have asked, you will discover how to get the most out of this *Learning Guide* as you progress through your algebra course.

We hope you have a great semester!

Bob Blitzer, *Miami-Dade College, FL*
C. Brad Davis

Great Questions that Students are Asking!

What is the relationship between this Learning Guide and the textbook?

Each section of the textbook begins with a list of learning objectives that focus on the section's most important ideas. Your *Learning Guide* is organized around the textbook's learning objectives. These objectives are the headers for the *Solved Problems* and *Pencil Problems* that form the essence of this guide.

What am I supposed to do with this Learning Guide? Why is it 3-hole punched?

- It is 3-hole punched so you can use it as the basis of your **course notebook**.
- You should **insert your lecture notes** at the beginning of each section.
- The main component of the *Learning Guide* is the Guided Practice. This is a series of ***Solved Problems*** that are paired with unsolved ***Pencil Problems*** for you to attempt. Answers to these problems are given at the end of each section.

What are the benefits of using this Learning Guide?

- It will help you become better organized. This includes organizing your class notes, assigned homework, quizzes, and tests.
- It will enable you to use your textbook more efficiently.
- It will help increase your study skills.
- It will help you to prepare for the chapter tests.

About this *Learning Guide* (continued)

What else should I know about <u>College Algebra</u>?

As you have noticed, math textbooks can be lengthy. It is unlikely that your course will cover all the material in the entire book, so be sure to pay attention to the syllabus provided by your instructor.

You will notice that this *Learning Guide* uses the same subdivisions as the main textbook. Each Objective is written out so you will know exactly what concept is presented in the various examples. This will allow you to easily find the specific help you need as you move from resource to resource, and will allow you to easily skip the topics that your instructor is not planning to cover.

To take this course you should have the following:

- ***College Algebra 6th edition*** textbook (or eBook)
- ***Learning Guide***

Additional resources available for this course (optional, unless specified by your instructor):

- ***Student Solution Manual*** (contains solutions to selected exercises)
- ***MyMathLab/MathXL*** (online homework, exercise tutorials, and more)

Note to Instructors:

Thank you for choosing this *Learning Guide* for your class. Its design and purpose is to facilitate the teaching and learning process by keeping your students organized. Please encourage your students to read these introductory pages.

We hope you and your class have a rewarding semester.

Bob Blitzer, *Miami-Dade College, FL*
C. Brad Davis

Section P.1
Algebraic Expressions, Mathematical Models, and Real Numbers

It costs how much?

You are looking ahead to the next school year and wondering how much money you will need.
Is there any way that you can use trends for college costs over the past few years to predict how much college will cost next year?
In the Exercise Set for this section, you will use a model that will allow you to project average costs at private U.S. colleges in the near future.

Objective #1: Evaluate algebraic expressions.

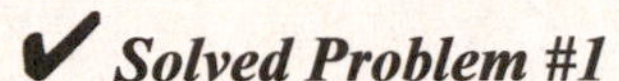

Solved Problem #1

1. Evaluate $8+6(x-3)^2$ for $x=13$.

$$\begin{aligned}8+6(x-3)^2 &= 8+6(13-3)^2\\ &= 8+6(10)^2\\ &= 8+6(100)\\ &= 8+600\\ &= 608\end{aligned}$$

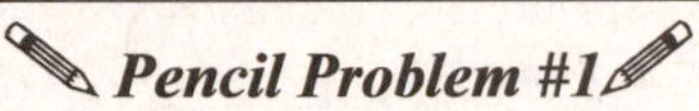

Pencil Problem #1

1. Evaluate $4+5(x-7)^3$ for $x=9$.

Objective #2: Use mathematical models.

Solved Problem #2

2. The formula $T=4x^2+341x+3194$ models the average cost of tuition and fees, T, for public U.S. colleges for the school year ending x years after 2000. Use this formula to project the average cost of tuition and fees at public U.S. colleges for the school year ending in 2015.

Because 2015 is 15 years after 2000, we substitute 15 for x in the formula.

$$T=4x^2+341x+3194$$
$$T=4(15)^2+341(15)+3194$$
$$T=4(225)^2+341(15)+3194$$
$$T=900+5115+3194$$
$$T=9209$$

The formula indicates that for the school year ending in 2015, the average cost of tuition and fees at public U.S. colleges will be \$9209.

Pencil Problem #2

2. The formula $T=26x^2+819x+15{,}527$ models the average cost of tuition and fees, T, for private U.S. colleges for the school year ending x years after 2000. Use this formula to project the average cost of tuition and fees at private U.S. colleges for the school year ending in 2013.

Objective #3: Find the intersection of two sets.

Solved Problem #3

3. Find the intersection: $\{3, 4, 5, 6, 7\} \cap \{3, 7, 8, 9\}$.

The elements common to $\{3, 4, 5, 6, 7\}$ and $\{3, 7, 8, 9\}$ are 3 and 7.

$\{3, 4, 5, 6, 7\} \cap \{3, 7, 8, 9\} = \{3, 7\}$

Pencil Problem #3

3. Find the intersection: $\{1, 2, 3, 4\} \cap \{2, 4, 5\}$.

Objective #4: Find the union of two sets.

Solved Problem #4

4. Find the union: $\{3, 4, 5, 6, 7\} \cup \{3, 7, 8, 9\}$.

List the elements from the first set: 3, 4, 5, 6, and 7. Now list any elements from the second set not in the first: 8 and 9.

$\{3, 4, 5, 6, 7\} \cup \{3, 7, 8, 9\} = \{3, 4, 5, 6, 7, 8, 9\}$

Pencil Problem #4

4. Find the union: $\{1, 2, 3, 4\} \cup \{2, 4, 5\}$.

Objective #5: Recognize subsets of the real numbers..

Solved Problem #5

5. Consider the following set of numbers:
$\left\{-9, -1.3, 0, 0.\overline{3}, \frac{\pi}{2}, \sqrt{9}, \sqrt{10}\right\}$.

5a. List the natural numbers.

The natural numbers are used for counting. The only natural number is $\sqrt{9}$ because $\sqrt{9} = 3$.

5b. List the rational numbers.

All numbers that can be expressed as quotients of integers are rational numbers: $-9\left(-9 = \frac{-9}{1}\right)$, $0\left(0 = \frac{0}{1}\right)$, and $\sqrt{9}\left(\sqrt{9} = \frac{3}{1}\right)$. All numbers that are terminating or repeating decimals are rational numbers: -1.3 and $0.\overline{3}$.

Pencil Problem #5

5. Consider the following set of numbers:
$\left\{-11, -\frac{5}{6}, 0, 0.75, \sqrt{5}, \pi, \sqrt{64}\right\}$.

5a. List the natural numbers.

5b. List the rational numbers.

Objective #6: Use inequality symbols.

✔ *Solved Problem #6*

6. Indicate whether each statement is true or false.

6a. $-8 > -3$

This statement is false. Because -8 lies to the left of -3 on a number line, -8 is less than -3. So, $-8 < -3$.

6b. $9 \le 9$

This statement is true because $9 = 9$.

Pencil Problem #6

6. Indicate whether each statement is true or false.

6a. $-7 < -2$

6b. $-5 \ge 2$

Objective #7: Evaluate absolute value.

✔ *Solved Problem #7*

7. Rewrite each expression without absolute value bars.

7a. $|1-\sqrt{2}|$

Because $\sqrt{2} \approx 1.4$, the number $1-\sqrt{2}$ is negative.
Thus, $|1-\sqrt{2}| = -(1-\sqrt{2}) = \sqrt{2}-1$.

7b. $|\pi - 3|$

Because $\pi \approx 3.14$, the number $\pi - 3$ is positive.
Thus, $|\pi - 3| = \pi - 3$.

7c. $\frac{|x|}{x}$ if $x > 0$

If $x > 0$, then $|x| = x$. Thus, $\frac{|x|}{x} = \frac{x}{x} = 1$.

Pencil Problem #7

7. Rewrite each expression without absolute value bars.

7a. $|12 - \pi|$

7b. $|\sqrt{2} - 5|$

7c. $\frac{-3}{|-3|}$

Objective #8: Use absolute value to express distance.

✔ *Solved Problem #8*

8. Find the distance between –4 and 5 on the real number line.

$|-4-5| = |-9| = 9$

Pencil Problem #8

8. Find the distance between –19 and –4 on the real number line.

Objective #9: Identify properties of the real numbers.

✔ *Solved Problem #9*

9. State the name of the property illustrated.

9a. $2+\sqrt{5}=\sqrt{5}+2$

The order of the numbers in the addition has changed. This illustrates the commutative property of addition.

9b. $1\cdot(x+3)=x+3$

One has been deleted from a product. This illustrates the identity property of multiplication.

Pencil Problem #9

9. State the name of the property illustrated.

9a. $6+(2+7)=(6+2)+7$

9b. $2(-8+6)=-16+12$

Objective #10: Simplify algebraic expressions.

✔ *Solved Problem #10*

10. Simplify: $6+4[7-(x-2)]$.

$$\begin{aligned} 6+4[7-(x-2)] &= 6+4[7-x+2] \\ &= 6+4[9-x] \\ &= 6+36-4x \\ &= (6+36)-4x \\ &= 42-4x \end{aligned}$$

Pencil Problem #10

10. Simplify: $7-4[3-(4y-5)]$.

Answers for Pencil Problems *(Textbook Exercise references in parentheses)*:

1. 44 *(P.1 #9)* **2.** $30,568 *(P.1 #131c)* **3.** $\{2, 4\}$ *(P.1 #21)* **4.** $\{1, 2, 3, 4, 5\}$ *(P.1 #29)*
5. a. $\sqrt{64}$ **b.** $-11, -\frac{5}{6}, 0, 0.75, \sqrt{64}$ *(P.1 #37)* **6. a.** true **b.** false **7. a.** $12-\pi$ *(P.1 #53)*
b. $5-\sqrt{2}$ *(P.1 #55)* **c.** -1 *(P.1 #57)* **8.** 15 *(P.1 #71)* **9. a.** associative property of addition *(P.1 #77)*
b. distributive property *(P.1 #81)* **10.** $16y-25$ *(P.1 #93)*

Section P.2
Exponents and Scientific Notation

WOW, THAT'S BIG!

Did you know that in the summer of 2012 the national debt passed \$16,000,000,000,000 or \$16 trillion? Yes, that's 12 zeros you count. In this section, you will express the national debt in a form called *scientific notation* and use this form to calculate your share of the debt.

Objective #1: Use the product rule.

✔ *Solved Problem #1*

1. Multiply using the product rule.

1a. $3^3 \cdot 3^2$

$3^3 \cdot 3^2 = 3^{3+2} = 3^5$ or 243

1b. $(4x^3y^4)(10x^2y^6)$

$(4x^3y^4)(10x^2y^6) = 4 \cdot 10 \cdot x^{3+2} \cdot y^{4+6} = 40x^5y^{10}$

Pencil Problem #1

1. Multiply using the product rule.

1a. $x^3 \cdot x^7$

1b. $(-9x^3y)(-2x^6y^4)$

Objective #2: Use the quotient rule.

✔ *Solved Problem #2*

2. Divide using the quotient rule.

2a. $\frac{(-3)^6}{(-3)^3}$

$\frac{(-3)^6}{(-3)^3} = (-3)^{6-3} = (-3)^3$ or -27

2b. $\frac{27x^{14}y^8}{3x^3y^5}$

$\frac{27x^{14}y^8}{3x^3y^5} = \frac{27}{3} \cdot x^{14-3} \cdot y^{8-5} = 9x^{11}y^3$

Pencil Problem #2

2. Divide using the quotient rule.

2a. $\frac{2^8}{2^4}$

2b. $\frac{25a^{13}b^4}{-5a^2b^3}$

Objective #3: Use the zero-exponent rule.

✔ *Solved Problem #3*

3. Evaluate -8^0.

Because there are no parentheses only 8 is raised to the 0 power.

$-8^0 = -(8^0) = -1$

Pencil Problem #3

3. Evaluate $(-3)^0$.

Objective #4: Use the negative-exponent rule.

✔ *Solved Problem #4*

4. Write with a positive exponent. Simplify, if possible.

4a. 5^{-2}

$5^{-2} = \frac{1}{5^2} = \frac{1}{25}$

4b. $3x^{-6}y^4$

$3x^{-6}y^4 = 3 \cdot \frac{1}{x^6} \cdot y^4 = \frac{3y^4}{x^6}$

Pencil Problem #4

4. Write with a positive exponent. Simplify, if possible.

4a. 4^{-3}

4b. $(4x^3)^{-2}$

Objective #5: Use the power rule.

✔ *Solved Problem #5*

5. Simplify using the power rule.

5a. $(3^3)^2$

$(3^3)^2 = 3^{3 \cdot 2} = 3^6$ or 729

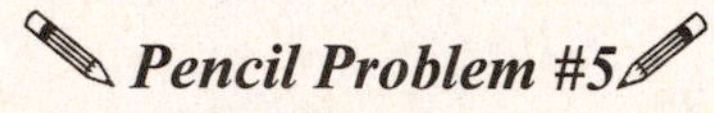

Pencil Problem #5

5. Simplify using the power rule.

5a. $(2^2)^3$

5b. $(y^7)^{-2}$

$(y^7)^{-2} = y^{7(-2)} = y^{-14} = \dfrac{1}{y^{14}}$

5b. $(x^{-5})^3$

Objective #6: Find the power of a product.

✔ *Solved Problem #6*

6. Simplify: $(-4x)^3$.

$(-4x)^3 = (-4)^3(x)^3 = -64x^3$

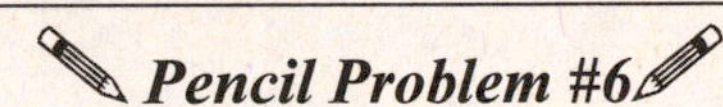

Pencil Problem #6

6. Simplify: $(8x^3)^2$.

Objective #7: Find the power of a quotient.

✔ *Solved Problem #7*

7. Simplify: $\left(-\dfrac{2}{y}\right)^5$.

$\left(-\dfrac{2}{y}\right)^5 = \dfrac{(-2)^5}{y^5} = \dfrac{-32}{y^5} = -\dfrac{32}{y^5}$

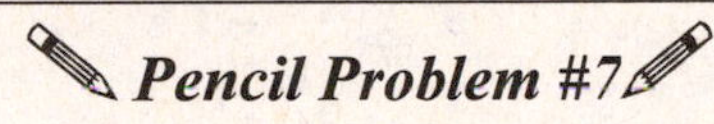

Pencil Problem #7

7. Simplify: $\left(-\dfrac{4}{x}\right)^3$.

Objective #8: Simplify exponential expressions.

✔ *Solved Problem #8*

8. Simplify.

8a. $(2x^3y^6)^4$

$(2x^3y^6)^4 = (2)^4(x^3)^4(y^6)^4$
$= 2^4 x^{3\cdot 4} y^{6\cdot 4}$
$= 16x^{12}y^{24}$

Pencil Problem #8

8. Simplify.

8a. $(-3x^2y^5)^2$

8b. $(-6x^2y^5)(3xy^3)$

$$(-6x^2y^5)(3xy^3) = (-6)(3)x^2xy^5y^3$$
$$= -18x^{2+1}y^{5+3}$$
$$= -18x^3y^8$$

8b. $(3x^4)(2x^7)$

8c. $\dfrac{100x^{12}y^2}{20x^{16}y^{-4}}$

$$\frac{100x^{12}y^2}{20x^{16}y^{-4}} = \left(\frac{100}{20}\right)\left(\frac{x^{12}}{x^{16}}\right)\left(\frac{y^2}{y^{-4}}\right)$$
$$= 5x^{12-16}y^{2-(-4)}$$
$$= 5x^{-4}y^6$$
$$= \frac{5y^6}{x^4}$$

8c. $\dfrac{24x^3y^5}{32x^7y^{-9}}$

8d. $\left(\dfrac{5x}{y^4}\right)^{-2}$

$$\left(\frac{5x}{y^4}\right)^{-2} = \frac{(5x)^{-2}}{(y^4)^{-2}}$$
$$= \frac{5^{-2}x^{-2}}{y^{-6}}$$
$$= \frac{y^6}{5^2x^2}$$
$$= \frac{y^6}{25x^2}$$

8d. $\left(\dfrac{5x^3}{y}\right)^{-2}$

Objective #9: Use scientific notation.

✔ Solved Problem #9

9. In 9a and 9b, write each number in decimal notation.

9a. -2.6×10^{9}

Move the decimal point 9 places to the right.
$-2.6\times10^{9}=-2,600,000,000$

9b. 3.017×10^{-6}

Move the decimal point 6 places to the left.
$3.017\times10^{-6}=0.000003017$

9c. In 9c and 9d, write each number in scientific notation.

5,210,000,000

The decimal point needs to be moved 9 places to the left.
$5,210,000,000=5.21\times10^{9}$

9d. −0.000000006893

The decimal point needs to be moved 8 places to the right.
$-0.00000006893=-6.893\times10^{-8}$

Pencil Problem #9

9. In 9a and 9b, write each number in decimal notation.

9a. -7.16×10^{6}

9b. 7.9×10^{-1}

9c. In 9c and 9d, write each number in scientific notation.

32,000

9b. −0.00000000504

9e. In 9e and 9f, perform the indicated computations. Write the answers in scientific notation.

$(7.1\times10^{5})(5\times10^{-7})$

$$(7.1\times10^{5})(5\times10^{-7}) = (7.1\times5)\times10^{5+(-7)}$$
$$= 35.5\times10^{-2}$$
$$= (3.55\times10^{1})\times10^{-2}$$
$$= 3.55\times10^{-1}$$

9f. $\dfrac{1.2\times10^{6}}{3\times10^{-3}}$

$$\frac{1.2\times10^{6}}{3\times10^{-3}} = \frac{1.2}{3}\times10^{6-(-3)}$$
$$= 0.4\times10^{9}$$
$$= (4\times10^{-1})\times10^{9}$$
$$= 4\times10^{8}$$

9e. In 9e and 9f, perform the indicated computations. Write the answers in scientific notation.

$(1.6\times10^{15})(4\times10^{-11})$

9f. $\dfrac{2.4\times10^{-2}}{4.8\times10^{-6}}$

Answers for Pencil Problems *(Textbook Exercise references in parentheses)*:

1a. x^{10} *(P.2 #27)* **1b.** $18x^{9}y^{5}$ *(P.2 #47)* **2a.** 16 *(P.2 #17)* **2b.** $-5a^{11}b$ *(P.2 #51)*

3. 1 *(P.2 #7)* **4a.** $\dfrac{1}{64}$ *(P.2 #11)* **4b.** $\dfrac{1}{16x^{6}}$ *(P.2 #55)*

5a. 64 *(P.2 #15)* **5b.** $\dfrac{1}{x^{15}}$ *(P.2 #33)*

6. $64x^{6}$ *(P.2 #39)* **7.** $-\dfrac{64}{x^{3}}$ *(P.2 #41)*

8a. $9x^{4}y^{10}$ *(P.2 #43)* **8b.** $6x^{11}$ *(P.2 #45)* **8c.** $\dfrac{3y^{14}}{4x^{4}}$ *(P.2 #57)* **8d.** $\dfrac{y^{2}}{25x^{6}}$ *(P.2 #59)*

9a. $-7{,}160{,}000$ *(P.2 #69)* **9b.** 0.79 *(P.2 #71)* **9c.** 3.2×10^{4} *(P.2 #77)* **9d.** -5.04×10^{-9} *(P.2 #85)*
9e. 6.4×10^{4} *(P.2 #89)* **9f.** 5×10^{3} *(P.2 #101)*

Section P.3
Radicals and Rational Exponents

Radicals in Space?

What does space travel have to do with radicals?

Imagine that in the future we will be able to travel at velocities approaching the speed of light
(approximately 186,000 miles per second). According to Einstein's theory of special relativity, time would pass more quickly on Earth than it would in the moving spaceship.

Objective #1: Evaluate square roots.

Solved Problem #1

1a. Evaluate $\sqrt{81}$.

$\sqrt{81} = 9$ Check: $9^2 = 81$

1b. Evaluate $-\sqrt{9}$.

$-\sqrt{9} = -3$ Check: $(-3)^2 = 9$

1c. Evaluate $\sqrt{\frac{1}{25}}$.

$\sqrt{\frac{1}{25}} = \frac{1}{5}$ Check: $\left(\frac{1}{5}\right)^2 = \frac{1}{25}$

1d. Evaluate $\sqrt{36+64}$.

$\sqrt{36+64} = \sqrt{100} = 10$

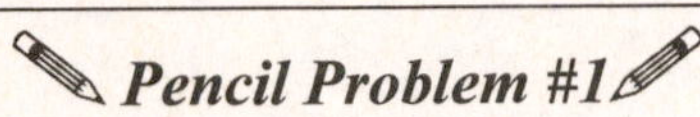

Pencil Problem #1

1a. Evaluate $\sqrt{36}$.

1b. Evaluate $-\sqrt{36}$.

1c. Evaluate $\sqrt{\frac{1}{81}}$.

1d. Evaluate $\sqrt{25-16}$.

1e. Evaluate $\sqrt{36}+\sqrt{64}$.

$\sqrt{36}+\sqrt{64}=6+8=14$

1e. Evaluate $\sqrt{25}-\sqrt{16}$.

Objective #2: Simplify expressions of the form $\sqrt{a^2}$.

✔ *Solved Problem #2*

2. Evaluate $\sqrt{(-6)^2}$.

$\sqrt{(-6)^2}=|-6|=6$

Pencil Problem #2

2. Evaluate $\sqrt{(-13)^2}$.

Objective #3: Use the product rule to simplify square roots.

✔ *Solved Problem #3*

3a. Simplify $\sqrt{75}$.

$\sqrt{75}=\sqrt{25\cdot 3}=\sqrt{25}\cdot\sqrt{3}=5\sqrt{3}$

3b. Simplify $\sqrt{5x}\cdot\sqrt{10x}$.

$$\begin{aligned}\sqrt{5x}\cdot\sqrt{10x}&=\sqrt{5x\cdot 10x}\\&=\sqrt{50x^2}\\&=\sqrt{25x^2\cdot 2}\\&=\sqrt{25x^2}\,\sqrt{2}\\&=5x\sqrt{2}\end{aligned}$$

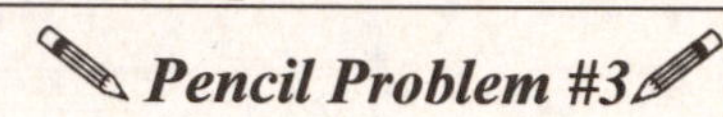

Pencil Problem #3

3a. Simplify $\sqrt{50}$.

3b. Simplify $\sqrt{2x}\cdot\sqrt{6x}$.

Objective #4: Use the quotient rule to simplify square roots.

✔ *Solved Problem #4*

4a. Simplify $\sqrt{\frac{25}{16}}$.

$$\sqrt{\frac{25}{16}} = \frac{\sqrt{25}}{\sqrt{16}} = \frac{5}{4}$$

4b. Simplify $\frac{\sqrt{150x^3}}{\sqrt{2x}}$.

$$\begin{aligned}\frac{\sqrt{150x^3}}{\sqrt{2x}} &= \sqrt{\frac{150x^3}{2x}} \\ &= \sqrt{75x^2} \\ &= \sqrt{25x^2 \cdot 3} \\ &= \sqrt{25x^2}\sqrt{3} \\ &= 5x\sqrt{3}\end{aligned}$$

Pencil Problem #4

4a. Simplify $\sqrt{\frac{49}{16}}$.

4b. Simplify $\frac{\sqrt{48x^3}}{\sqrt{3x}}$.

Objective #5: Add and subtract square roots.

✔ *Solved Problem #5*

5a. Add: $8\sqrt{13} + 9\sqrt{13}$.

$$8\sqrt{13} + 9\sqrt{13} = (8+9)\sqrt{13} = 17\sqrt{13}$$

Pencil Problem #5

5a. Subtract: $6\sqrt{17x} - 8\sqrt{17x}$.

5b. Subtract: $6\sqrt{18x}-4\sqrt{8x}$.

$$\begin{aligned}6\sqrt{18x}-4\sqrt{8x} &= 6\sqrt{9\cdot 2x}-4\sqrt{4\cdot 2x}\\ &= 6\cdot 3\sqrt{2x}-4\cdot 2\sqrt{2x}\\ &= 18\sqrt{2x}-8\sqrt{2x}\\ &= (18-8)\sqrt{2x}\\ &= 10\sqrt{2x}\end{aligned}$$

5b. Add: $3\sqrt{18}+5\sqrt{50}$.

Objective #6: Rationalize denominators.

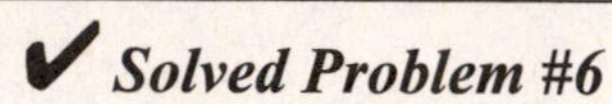

6a. Rationalize the denominator : $\dfrac{6}{\sqrt{12}}$.

Multiply by $\sqrt{3}$ to obtain the square root of a perfect square, $\sqrt{12}\cdot\sqrt{3}=\sqrt{36}$.

$$\frac{6}{\sqrt{12}}\cdot\frac{\sqrt{3}}{\sqrt{3}}=\frac{6\sqrt{3}}{\sqrt{36}}=\frac{6\sqrt{3}}{6}=\sqrt{3}$$

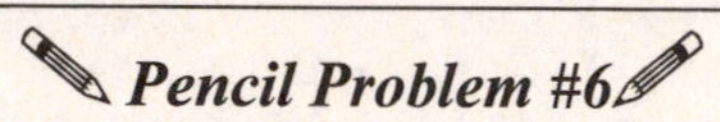

6a. Rationalize the denominator : $\dfrac{\sqrt{2}}{\sqrt{5}}$.

6b. Rationalize the denominator : $\dfrac{8}{4+\sqrt{5}}$.

Multiply by $4-\sqrt{5}$, the conjugate of $4+\sqrt{5}$.

$$\begin{aligned}\frac{8}{4+\sqrt{5}}\cdot\frac{4-\sqrt{5}}{4-\sqrt{5}} &= \frac{8(4-\sqrt{5})}{4^2-(\sqrt{5})^2}\\ &= \frac{8(4-\sqrt{5})}{16-5}\\ &= \frac{8(4-\sqrt{5})}{11} \text{ or } \frac{32-8\sqrt{5}}{11}\end{aligned}$$

6b. Rationalize the denominator : $\dfrac{7}{\sqrt{5}-2}$.

Objective #7: Evaluate and perform operations with higher roots.

✔ Solved Problem #7

7a. Simplify $\sqrt[5]{8}\cdot\sqrt[5]{8}$.

$$\begin{aligned}\sqrt[5]{8}\cdot\sqrt[5]{8} &= \sqrt[5]{8\cdot 8}\\ &= \sqrt[5]{64}\\ &= \sqrt[5]{32\cdot 2}\\ &= \sqrt[5]{32}\cdot\sqrt[5]{2}\\ &= 2\sqrt[5]{2}\end{aligned}$$

7b. Simplify $\sqrt[3]{\dfrac{125}{27}}$.

$$\sqrt[3]{\frac{125}{27}} = \frac{\sqrt[3]{125}}{\sqrt[3]{27}} = \frac{5}{3}$$

7c. Subtract: $3\sqrt[3]{81} - 4\sqrt[3]{3}$.

$$\begin{aligned}3\sqrt[3]{81} - 4\sqrt[3]{3} &= 3\sqrt[3]{27\cdot 3} - 4\sqrt[3]{3}\\ &= 3\cdot 3\sqrt[3]{3} - 4\sqrt[3]{3}\\ &= 9\sqrt[3]{3} - 4\sqrt[3]{3}\\ &= (9-4)\sqrt[3]{3}\\ &= 5\sqrt[3]{3}\end{aligned}$$

Pencil Problem #7

7a. Simplify $\sqrt[3]{9}\cdot\sqrt[3]{6}$.

7b. Simplify $\dfrac{\sqrt[5]{64x^6}}{\sqrt[5]{2x}}$.

7c. Add: $5\sqrt[3]{16} + \sqrt[3]{54}$.

Objective #8: Understand and use rational exponents..

Solved Problem #8

8a. Simplify $(-8)^{\frac{1}{3}}$.

$(-8)^{\frac{1}{3}} = \sqrt[3]{-8} = -2$

8b. Simplify $32^{-\frac{2}{5}}$.

$32^{-\frac{2}{5}} = \dfrac{1}{32^{\frac{2}{5}}} = \dfrac{1}{(\sqrt[5]{32})^2} = \dfrac{1}{2^2} = \dfrac{1}{4}$

8c. Simplify $\dfrac{20x^4}{5x^{\frac{3}{2}}}$.

$\dfrac{20x^4}{5x^{\frac{3}{2}}} = \dfrac{20}{5} \cdot x^{4-\frac{3}{2}} = 4x^{\frac{8}{2}-\frac{3}{2}} = 4x^{\frac{5}{2}}$

8d. Simplify $\sqrt[6]{x^3}$.

$\sqrt[6]{x^3} = x^{\frac{3}{6}} = x^{\frac{1}{2}} = \sqrt{x}$

Pencil Problem #8

8a. Simplify $36^{\frac{1}{2}}$.

8b. Simplify $125^{\frac{2}{3}}$.

8c. Simplify $(7x^{\frac{1}{3}})(2x^{\frac{1}{4}})$.

8d. Simplify $\sqrt[6]{x^4}$.

Answers for Pencil Problems *(Textbook Exercise references in parentheses)*:

1a. 6 *(P.3 #1)* **1b.** −6 *(P.3 #3)* **1c.** $\frac{1}{9}$ *(P.3 #23)* **1d.** 3 *(P.3 #7)* **1e.** 1 *(P.3 #9)*

2. 13 *(P.3 #11)* **3a.** $5\sqrt{2}$ *(P.3 #13)* **3b.** $2x\sqrt{3}$ *(P.3 #17)*

4a. $\frac{7}{4}$ *(P.3 #23)* **4b.** $4x$ *(P.3 #27)* **5a.** $-2\sqrt{17x}$ *(P.3 #35)* **5b.** $34\sqrt{2}$ *(P.3 #41)*

6a. $\frac{\sqrt{10}}{5}$ *(P.3 #47)* **6b.** $7(\sqrt{5}+2)$ *(P.3 #51)*

7a. $3\sqrt[3]{2}$ *(P.3 #71)* **7b.** $2x$ *(P.3 #73)* **7c.** $13\sqrt[3]{2}$ *(P.3 #77)*

8a. 6 *(P.3 #83)* **8b.** 25 *(P.3 #87)* **8c.** $14x^{\frac{7}{12}}$ *(P.3 #91)* **8d.** $\sqrt[3]{x^2}$ *(P.3 #105)*

Section P.4
Polynomials

What Are the Best Dimensions for a Box?

Many children get excited about gift boxes of all shapes and sizes, with the possible *exception* of clothing-sized boxes. (I must confess I dreaded boxes of that size.)

While completing the application exercises in this section of the textbook, we will use polynomials to model the dimensions of a box. We will then apply the concepts of this section to model the area of the box's base and its volume.

Objective #1: Understand the vocabulary of polynomials.

✔ *Solved Problem #1*

1. True or false: $7x^5 - 3x^3 + 8$ is a polynomial of degree 7 with three terms.

False. The expression $7x^5 - 3x^3 + 8$ is a polynomial with three terms, but its degree is 5, not 7.

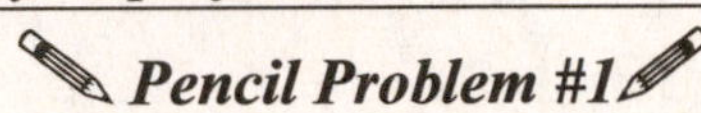

Pencil Problem #1

1. True or false: $x^2 - 4x^3 + 9x - 12x^4 + 63$ is a polynomial of degree 2 with five terms.

Objective #2: Add and subtract polynomials.

✔ *Solved Problem #2*

2a. Add: $(-17x^3 + 4x^2 - 11x - 5) + (16x^3 - 3x^2 + 3x - 15)$.

$(-17x^3 + 4x^2 - 11x - 5) + (16x^3 - 3x^2 + 3x - 15)$
$= (-17x^3 + 16x^3) + (4x^2 - 3x^2) + (-11x + 3x) + (-5 - 15)$
$= -x^3 + x^2 + (-8x) + (-20)$
$= -x^3 + x^2 - 8x - 20$

Pencil Problem #2

2a. Add: $(-6x^3 + 5x^2 - 8x + 9) + (17x^3 + 2x^2 - 4x - 13)$.

2b. Subtract:
$(13x^3 - 9x^2 - 7x + 1) - (-7x^3 + 2x^2 - 5x + 9)$.

$(13x^3 - 9x^2 - 7x + 1) - (-7x^3 + 2x^2 - 5x + 9)$
$= (13x^3 - 9x^2 - 7x + 1) + (7x^3 - 2x^2 + 5x - 9)$
$= (13x^3 + 7x^3) + (-9x^2 - 2x^2) + (-7x + 5x) + (1 - 9)$
$= 20x^3 + (-11x^2) + (-2x) + (-8)$
$= 20x^3 - 11x^2 - 2x - 8$

2b. Subtract:
$(17x^3 - 5x^2 + 4x - 3) - (5x^3 - 9x^2 - 8x + 11)$.

Objective #3: Multiply polynomials.

✔ *Solved Problem #3*

3. Multiply: $(5x - 2)(3x^2 - 5x + 4)$.

$(5x - 2)(3x^2 - 5x + 4)$
$= 5x(3x^2 - 5x + 4) - 2(3x^2 - 5x + 4)$
$= 5x \cdot 3x^2 + 5x(-5x) + 5x \cdot 4 - 2 \cdot 3x^2 - 2(-5x) - 2 \cdot 4$
$= 15x^3 - 25x^2 + 20x - 6x^2 + 10x - 8$
$= 15x^3 - 31x^2 + 30x - 8$

Pencil Problem #3

3. Multiply: $(2x - 3)(x^2 - 3x + 5)$.

Objective #4: Use FOIL in polynomial multiplication.

✔ *Solved Problem #4*

4. Multiply: $(7x - 5)(4x - 3)$.
Use FOIL.

First: $7x \cdot 4x$ Outside: $7x(-3)$
Inside: $-5 \cdot 4x$ Last: $-5(-3)$

$(7x - 5)(4x - 3)$
$= 7x \cdot 4x + 7x(-3) - 5 \cdot 4x - 5(-3)$
$= 28x^2 - 21x - 20x + 15$
$= 28x^2 - 41x + 15$

Pencil Problem #4

4. Multiply: $(3x + 5)(2x + 1)$.

Objective #5: Use special products in polynomial multiplication.

✔ Solved Problem #5

5a. Multiply: $(7x+8)(7x-8)$.

Use $(A+B)(A-B)=A^2-B^2$.

$$(7x+8)(7x-8)=(7x)^2-8^2$$
$$=49x^2-64$$

5b. Multiply: $(5x+4)^2$.

Use $(A+B)^2=A^2+2AB+B^2$.

$$(5x+4)^2=(5x)^2+2(5x)(4)+4^2$$
$$=25x^2+40x+16$$

5c. Multiply: $(x-9)^2$.

Use $(A-B)^2=A^2-2AB+B^2$.

$$(x-9)^2=x^2-2\cdot x\cdot 9+9^2$$
$$=x^2-18x+81$$

Pencil Problem #5

5a. Multiply: $(5-7x)(5+7x)$.

5b. Multiply: $(2x+3)^2$.

5c. Multiply: $(x-3)^2$.

Objective #6: Perform operations with polynomials in several variables.

✔ Solved Problem #6

6a. Subtract:
$(x^3 - 4x^2y + 5xy^2 - y^3) - (x^3 - 6x^2y + y^3)$.

$$\begin{aligned}&(x^3 - 4x^2y + 5xy^2 - y^3) - (x^3 - 6x^2y + y^3)\\ &= (x^3 - 4x^2y + 5xy^2 - y^3) + (-x^3 + 6x^2y - y^3)\\ &= (x^3 - x^3) + (-4x^2y + 6x^2y) + 5xy^2 + (-y^3 - y^3)\\ &= 2x^2y + 5xy^2 - 2y^3\end{aligned}$$

6b. Multiply: $(7x - 6y)(3x - y)$.

Use FOIL.

$$\begin{aligned}&(7x - 6y)(3x - y)\\ &= 7x \cdot 3x + 7x(-y) - 6y \cdot 3x - 6y(-y)\\ &= 21x^2 - 7xy - 18xy + 6y^2\\ &= 21x^2 - 25xy + 6y^2\end{aligned}$$

Pencil Problem #6

6a. Add: $(4x^2y + 8xy + 11) + (-2x^2y + 5xy + 2)$.

6b. Multiply: $(7x + 5y)^2$.

Answers for Pencil Problems *(Textbook Exercise references in parentheses)*:

1. False *(P.4 #7)* **2a.** $11x^3 + 7x^2 - 12x - 4$ *(P.4 #9)* **2b.** $12x^3 + 4x^2 + 12x - 14$ *(P.4 #11)*
3. $2x^3 - 9x^2 + 19x - 15$ *(P.4 #17)* **4.** $6x^2 + 13x + 5$ *(P.4 #23)*
5a. $25 - 49x^2$ *(P.4 #35)* **5b.** $4x^2 + 12x + 9$ *(P.4 #43)* **5c.** $x^2 - 6x + 9$ *(P.4 #45)*
6a. $2x^2y + 13xy + 13$ *(P.4 #61)* **6b.** $49x^2 + 70xy + 25y^2$ *(P.4 #73)*

Section P.5
Factoring Polynomials

What's the sales price?

Many times retailers advertise their discounts in terms of percentages by which the price is reduced, such as 30% off. If a product still doesn't sell, the retailer may offer an additional 30% off the price that has already been reduced by 30%.

In this section's Exercise Set, you will see how the 30% discount followed by another 30% discount can be expressed as a polynomial. By factoring the polynomial and simplifying, you will see that our double discount means that we pay 49% of the original price.

Objective #1: Factor out the greatest common factor.

✔ *Solved Problem #1*

1a. Factor $10x^3 - 4x^2$.

2 is the greatest integer that divides 10 and 4. x^2 is the greatest expression that divides x^3 and x^2. The GCF is $2x^2$.

$$10x^3 - 4x^2 = 2x^2(5x) - 2x^2(2)$$
$$= 2x^2(5x - 2)$$

1b. Factor $2x(x-7)+3(x-7)$.

The GCF is the binomial factor $(x - 7)$.

$$2x(x-7)+3(x-7) = (x-7)(2x+3)$$

Pencil Problem #1

1a. Factor $3x^2 + 6x$.

1b. Factor $x(x+5)+3(x+5)$.

Objective #2: Factor by grouping.

✔ *Solved Problem #2*

2. Factor $x^3+5x^2-2x-10$.

The GCF of the first two terms is x^2, and the GCF of the last two terms is -2. After factoring out these GCFs, factor out the common binomial factor.

$$\begin{aligned} x^3+5x^2-2x-10 &= (x^3+5x^2)+(-2x-10) \\ &= x^2(x+5)-2(x+5) \\ &= (x+5)(x^2-2) \end{aligned}$$

Pencil Problem #2

2. Factor $x^3-2x^2+5x-10$.

Objective #3: Factor trinomials.

✔ *Solved Problem #3*

3a. Factor $x^2-5x-14$.

The leading coefficient is 1. We look for factors of -14 that sum to -5.

$-7(2)=-14$ and $-7+2=-5$
The numbers are -7 and 2.
$x^2-5x-14=(x-7)(x+2)$

Pencil Problem #3

3a. Factor $x^2-8x+15$.

3b. Factor $6x^2+19x-7$.

The leading coefficient is 6, not 1. $6x^2$ factors as $6x(x)$ or $3x(2x)$. -7 factors as $-7(1)$ or $7(-1)$.

The possible factorizations are

$(6x-7)(x+1)$	$(6x+1)(x-7)$
$(6x+7)(x-1)$	$(6x-1)(x+7)$
$(3x-7)(2x+1)$	$(3x+1)(2x-7)$
$(3x+7)(2x-1)$	$(3x-1)(2x+7)$

We want the combination, if there is one, that results in a sum of Outside and Inside terms of $19x$. Compute the sums of the Outside and Inside terms in the possible factorizations until you find one that results in $19x$.

For $(3x-1)(2x+7)$:
Outside: $3x(7)=21x$
Inside: $-1(2x)=-2x$
Sum: $21x+(-2x)=19x$
So, $6x^2+19x-7=(3x-1)(2x+7)$.

3b. Factor $9x^2-9x+2$.

Objective #4: Factor the difference of squares.

✔ *Solved Problem #4*

4. Factor: $36x^2-25$.

Note that $36x^2=(6x)^2$ and $25=5^2$ can both be expressed as squares.

Use $A^2-B^2=(A+B)(A-B)$.

$$36x^2-25=(6x)^2-5^2$$
$$=(6x+5)(6x-5)$$

Pencil Problem #4

4. Factor $9x^2-25y^2$.

Objective #5: Factor perfect square trinomials.

✔ *Solved Problem #5*

5a. Factor $x^2+14x+49$.

Note that the first term is the square of x, the last term is the square of 7, and the middle term is twice the product of x and 7.

Factor using $A^2+2AB+B^2=(A+B)^2$.

$$x^2+14x+49=x^2+2\cdot x\cdot 7+7^2$$
$$=(x+7)^2$$

5b. Factor $16x^2-56x+49$.

Note that the first term is the square of $4x$, the last term is the square of 7, and the middle term is twice the product of $4x$ and 7.

Factor using $A^2-2AB+B^2=(A-B)^2$.

$$16x^2-56x+49=(4x)^2-2\cdot 4x\cdot 7+7^2$$
$$=(4x-7)^2$$

Pencil Problem #5

5a. Factor x^2+2x+1.

5b. Factor $9x^2-6x+1$.

Objective #6: Factor the sum or difference of two cubes.

✔ *Solved Problem #6*

6a. Factor x^3+1.

Note that both terms can be expressed as cubes.

Factor using $A^3+B^3=(A+B)(A^2-AB+B^2)$.

$$x^3+1=x^3+1^3$$
$$=(x+1)(x^2-x\cdot 1+1^2)$$
$$=(x+1)(x^2-x+1)$$

Pencil Problem #6

6a. Factor x^3+27.

6b. Factor $125x^3 - 8$.

Note that the first term is the cube of $5x$ and the second term is the cube of 2.

Factor using $A^3 - B^3 = (A-B)(A^2 + AB + B^2)$.

$$\begin{aligned} 125x^3 - 8 &= (5x)^3 - 2^3 \\ &= (5x-2)[(5x)^2 + 5x \cdot 2 + 2^2] \\ &= (5x-2)(25x^2 + 10x + 4) \end{aligned}$$

6b. Factor $8x^3 - 1$.

Objective #7: Use a general strategy for factoring polynomials..

✔ *Solved Problem #7*

7a. Factor $3x^3 - 30x^2 + 75x$.

First, factor out the GCF, $3x$.

$3x^3 - 30x^2 + 75x = 3x(x^2 - 10x + 25)$

Now factor the trinomial. Find factors of 25 that sum to –10, or use the formula for a perfect square trinomial, $A^2 - 2AB + B^2 = (A-B)^2$.

$$\begin{aligned} 3x^3 - 30x^2 + 75x &= 3x(x^2 - 10x + 25) \\ &= 3x(x^2 - 2 \cdot x \cdot 5 + 5^2) \\ &= 3x(x-5)^2 \end{aligned}$$

7b. Factor $x^2 - 36a^2 + 20x + 100$.

Regroup factors and look for opportunities to factor within groupings.

$x^2 - 36a^2 + 20x + 100 = (x^2 + 20x + 100) - 36a^2$

Factor the expression in parentheses using $A^2 + 2AB + B^2 = (A+B)^2$.

$$\begin{aligned} (x^2 + 20x + 100) - 36a^2 &= (x^2 + 2 \cdot x \cdot 10 + 10^2) - 36a^2 \\ &= (x+10)^2 - 36a^2 \end{aligned}$$

Pencil Problem #7

7a. Factor $20y^4 - 45y^2$.

7b. Factor $x^2 - 12x + 36 - 49y^2$.

This last form is the difference of squares. Factor using $A^2 - B^2 = (A+B)(A-B)$.

$$\begin{aligned}(x+10)^2 - 36a^2 &= (x+10)^2 - (6a)^2 \\ &= [(x+10)+6a][(x+10)-6a] \\ &= (x+10+6a)(x+10-6a)\end{aligned}$$

So, $x^2 - 36a^2 + 20x + 100 = (x+10+6a)(x+10-6a)$.

Objective #8: Factor algebraic expressions containing fractional and negative exponents.

✔ Solved Problem #8

8. Factor and simplify: $x(x-1)^{-\frac{1}{2}} + (x-1)^{\frac{1}{2}}$.

The GCF is $(x-1)$ with the smaller exponent. Thus, the GCF is $(x-1)^{-\frac{1}{2}}$.

$$\begin{aligned}x(x-1)^{-\frac{1}{2}} + (x-1)^{\frac{1}{2}} &= (x-1)^{-\frac{1}{2}} \cdot x + (x-1)^{-\frac{1}{2}} \cdot (x-1) \\ &= (x-1)^{-\frac{1}{2}}[x+(x-1)] \\ &= (x-1)^{-\frac{1}{2}}(2x-1) \\ &= \frac{2x-1}{(x-1)^{1/2}}\end{aligned}$$

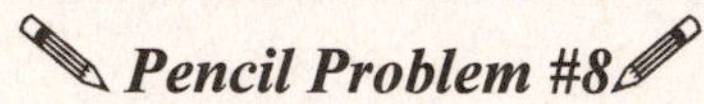

Pencil Problem #8

8. Factor and simplify: $(x+3)^{\frac{1}{2}} - (x+3)^{\frac{3}{2}}$.

Answers for Pencil Problems *(Textbook Exercise references in parentheses)*:

1a. $3x(x+2)$ *(P.5 #3)* **1b.** $(x+5)(x+3)$ *(P.5 #7)* **2.** $(x-2)(x^2+5)$ *(P.5 #11)*

3a. $(x-5)(x-3)$ *(P.5 #21)* **3b.** $(3x-2)(3x-1)$ *(P.5 #31)*

4. $(3x-5y)(3x+5y)$ *(P.5 #43)* **5a.** $(x+1)^2$ *(P.5 #49)* **5b.** $(3x-1)^2$ *(P.5 #55)*

6a. $(x+3)(x^2-3x+9)$ *(P.5 #57)* **6b.** $(2x-1)(4x^2+2x+1)$ *(P.5 #61)*

7a. $5y^2(2y+3)(2y-3)$ *(P.5 #83)* **7b.** $(x-6+7y)(x-6-7y)$ *(P.5 #85)* **8.** $-(x+3)^{\frac{1}{2}}(x+2)$ *(P.5 #97)*

Section P.6
Rational Expressions

Ouch! That Hurts!!

Though it may not be fun to get a flu shot, it is a great way to protect yourself from getting sick!

In this section of the textbook, one of the application problems will explore the costs for inoculating various percentages of the population.

Objective #1: Specify numbers that must be excluded from the domain of a rational expression.

✔ *Solved Problem #1*

1. Find all real numbers that must be excluded from the domain of each rational expression.

1a. $\frac{7}{x+5}$

The denominator, $x + 5$, would be 0 if $x = -5$. We must exclude -5 from the domain.

1b. $\frac{7x}{x^2-5x-14}$

Factor the denominator.

$x^2-5x-14=(x-7)(x+2)$

The first factor would be 0 if $x = 7$. The second factor would be 0 if $x = -2$. We must exclude -2 and 7 from the domain.

Pencil Problem #1

1. Find all real numbers that must be excluded from the domain of each rational expression.

1a. $\frac{7}{x-3}$

1b. $\frac{x+5}{x^2-25}$

Objective #2: Simplify rational expressions.

✔ *Solved Problem #2*

2a. Simplify $\dfrac{x^3+3x^2}{x+3}$.

Note that $x \neq -3$ since -3 would make the denominator 0.

Factor the numerator and divide out common factors.

$$\frac{x^3+3x^2}{x+3} = \frac{x^2(x+3)}{x+3} = \frac{x^2\cancel{(x+3)}}{\cancel{x+3}}$$
$$= x^2,\ x \neq -3$$

2b. Simplify $\dfrac{x^2-1}{x^2+2x+1}$.

By factoring the denominator, $x^2+2x+1=(x+1)^2$, we see that $x \neq -1$.

Factor the numerator and denominator and divide out common factors.

$$\frac{x^2-1}{x^2+2x+1} = \frac{(x+1)(x-1)}{(x+1)(x+1)} = \frac{\cancel{(x+1)}(x-1)}{\cancel{(x+1)}(x+1)}$$
$$= \frac{x-1}{x+1},\ x \neq -1$$

Pencil Problem #2

2a. Simplify $\dfrac{3x-9}{x^2-6x+9}$.

2b. Simplify $\dfrac{y^2+7y-18}{y^2-3y+2}$.

Objective #3: Multiply rational expressions.

✔ *Solved Problem #3*

3. Multiply: $\dfrac{x+3}{x^2-4}\cdot\dfrac{x^2-x-6}{x^2+6x+9}$.

Factor and divide by common factors.

$$\frac{x+3}{x^2-4}\cdot\frac{x^2-x-6}{x^2+6x+9} = \frac{x+3}{(x+2)(x-2)}\cdot\frac{(x-3)(x+2)}{(x+3)(x+3)}$$
$$= \frac{\cancel{x+3}}{\cancel{(x+2)}(x-2)}\cdot\frac{(x-3)\cancel{(x+2)}}{\cancel{(x+3)}(x+3)}$$
$$= \frac{x-3}{(x-2)(x+3)},\ x \neq -3,-2,2$$

To see which values must be excluded from the domain, look at the factored forms of the denominators in the second step.

Pencil Problem #3

3. Multiply: $\dfrac{x^2-5x+6}{x^2-2x-3}\cdot\dfrac{x^2-1}{x^2-4}$.

Objective #4: Divide rational expressions.

✔ Solved Problem #4

4. Divide: $\dfrac{x^2-2x+1}{x^3+x} \div \dfrac{x^2+x-2}{3x^2+3}$.

Invert the divisor and multiply.

$$\frac{x^2-2x+1}{x^3+x} \div \frac{x^2+x-2}{3x^2+3} = \frac{x^2-2x+1}{x^3+x} \cdot \frac{3x^2+3}{x^2+x-2}$$

$$= \frac{(x-1)\cancel{(x-1)}}{x\cancel{(x^2+1)}} \cdot \frac{3\cancel{(x^2+1)}}{(x+2)\cancel{(x-1)}}$$

$$= \frac{3(x-1)}{x(x+2)},\ x \neq -2, 0, 1$$

Pencil Problem #4

4. Divide: $\dfrac{x^2-25}{2x-2} \div \dfrac{x^2+10x+25}{x^2+4x-5}$.

Objective #5: Add and subtract rational expressions.

✔ Solved Problem #5

5a. Subtract: $\dfrac{x}{x+1} - \dfrac{3x+2}{x+1}$.

The expressions have the same denominator. Subtract numerators.

$$\frac{x}{x+1} - \frac{3x+2}{x+1} = \frac{x-(3x+2)}{x+1}$$

$$= \frac{x-3x-2}{x+1}$$

$$= \frac{-2x-2}{x+1}$$

$$= \frac{-2\cancel{(x+1)}}{\cancel{x+1}}$$

$$= -2,\ x \neq -1$$

Pencil Problem #5

5a. Add: $\dfrac{4x+1}{6x+5} + \dfrac{8x+9}{6x+5}$.

5b. Add: $\frac{3}{x+1}+\frac{5}{x-1}$.

The denominators are not equal and have no common factor. Use the property $\frac{a}{b}+\frac{c}{d}=\frac{ad+bc}{bd}$.

$$\frac{3}{x+1}+\frac{5}{x-1}=\frac{3(x-1)+5(x+1)}{(x+1)(x-1)}$$
$$=\frac{3x-3+5x+5}{(x+1)(x-1)}$$
$$=\frac{8x+2}{(x+1)(x-1)}$$
$$=\frac{2(4x+1)}{(x+1)(x-1)},\ x\neq -1,1$$

5b. Add: $\frac{3}{x+4}+\frac{6}{x+5}$.

5c. Subtract: $\frac{x}{x^2-10x+25}-\frac{x-4}{2x-10}$.

Factor the denominators.

$x^2-10x+25=(x-5)(x-5)$

$2x-10=2(x-5)$

LCD $=2(x-5)(x-5)$

$$\frac{x}{x^2-10x+25}-\frac{x-4}{2x-10}$$
$$=\frac{x}{(x-5)(x-5)}-\frac{x-4}{2(x-5)}$$
$$=\frac{2x}{2(x-5)(x-5)}-\frac{(x-4)(x-5)}{2(x-5)(x-5)}$$
$$=\frac{2x-(x-4)(x-5)}{2(x-5)^2}$$
$$=\frac{2x-(x^2-9x+20)}{2(x-5)^2}$$
$$=\frac{2x-x^2+9x-20}{2(x-5)^2}$$
$$=\frac{-x^2+11x-20}{2(x-5)^2},\ x\neq 5$$

5c. Subtract: $\frac{3x}{x^2+3x-10}-\frac{2x}{x^2+x-6}$.

Objective #6: Simplify complex rational expressions.

✔ *Solved Problem #6*

6a. Simplify: $\dfrac{\frac{1}{x}-\frac{3}{2}}{\frac{1}{x}+\frac{3}{4}}$.

Subtract and add in the numerator and denominator to obtain a single rational expression in each.

$$\frac{1}{x}-\frac{3}{2}=\frac{1\cdot 2}{x\cdot 2}-\frac{3\cdot x}{2\cdot x}=\frac{2}{2x}-\frac{3x}{2x}=\frac{2-3x}{2x}$$

$$\frac{1}{x}+\frac{3}{4}=\frac{1\cdot 4}{x\cdot 4}+\frac{3\cdot x}{4\cdot x}=\frac{4}{4x}+\frac{3x}{4x}=\frac{4+3x}{4x}$$

Now return to the original complex fraction.

$$\frac{\frac{1}{x}-\frac{3}{2}}{\frac{1}{x}+\frac{3}{4}}=\frac{\frac{2-3x}{2x}}{\frac{4+3x}{4x}}$$

$$=\frac{2-3x}{2x}\cdot\frac{4x}{4+3x}$$

$$=\frac{2-3x}{\cancel{2x}}\cdot\frac{2\cdot\cancel{2x}}{4+3x}$$

$$=\frac{2(2-3x)}{4+3x},\ x\neq 0,-\frac{4}{3}$$

6b. Simplify: $\dfrac{\frac{1}{x+7}-\frac{1}{x}}{7}$.

The LCD of the fractions within the complex fraction is $x(x+7)$. Multiply the numerator and the denominator of the complex fraction by the LCD.

$$\frac{\frac{1}{x+7}-\frac{1}{x}}{7}=\frac{\left(\frac{1}{x+7}-\frac{1}{x}\right)x(x+7)}{7x(x+7)}$$

$$=\frac{\frac{1}{x+7}\cdot x(x+7)-\frac{1}{x}\cdot x(x+7)}{7x(x+7)}$$

$$=\frac{x-(x+7)}{7x(x+7)}$$

$$=\frac{-7}{7x(x+7)}$$

$$=\frac{-1}{x(x+7)},\ x\neq -7,0$$

Pencil Problem #6

6a. Simplify: $\dfrac{1+\frac{1}{x}}{3-\frac{1}{x}}$.

6b. Simplify: $\dfrac{\frac{3}{x-2}-\frac{4}{x+2}}{\frac{7}{x^2-4}}$.

Answers for Pencil Problems *(Textbook Exercise references in parentheses)*:

1a. 3 *(P.6 #1)* **1b.** $-5, 5$ *(P.6 #3)* **2a.** $\frac{3}{x-3}, x \neq 3$ *(P.6 #7)* **2b.** $\frac{y+9}{y-1}, y \neq 1,2$ *(P.6 #11)*

3. $\frac{x-1}{x+2}, x \neq -2,-1,2,3$ *(P.6 #19)* **4.** $\frac{x-5}{2}, x \neq -5,1$ *(P.6 #29)*

5a. $2, x \neq -\frac{5}{6}$ *(P.6 #33)* **5b.** $\frac{9x+39}{(x+4)(x+5)}, x \neq -5,-4$ *(P.6 #41)*

5c. $\frac{x^2-x}{(x+5)(x-2)(x+3)}, x \neq -5,-3,2$ *(P.6 #53)*

6a. $\frac{x+1}{3x-1}, x \neq 0,\frac{1}{3}$ *(P.6 #61)* **6b.** $-\frac{x-14}{7}, x \neq -2,2$ *(P.6 #67)*

Section 1.1
Graphs and Graphing Utilities

Let it snow! Let it snow! Let it snow!

The arrival of snow can range from light flurries to a full-fledged blizzard. Snow can be welcomed as a beautiful backdrop to outdoor activities or it can be a nuisance and endanger drivers.

We will look at how graphs can be used to explain both mathematical concepts and everyday situations. Specifically, in the application exercises of this section of the textbook, you will match stories of varying snowfalls to the graphs that explain them.

Objective #1: Plot points in the rectangular coordinate system.

✔ Solved Problem #1

1a. Plot the points:
$A(-2, 4)$, $B(4, -2)$, $C(-3, 0)$, and $D(0, -3)$.

From the origin, point A is left 2 units and up 4 units.

From the origin, point B is right 4 units and down 2 units.

From the origin, point C is left 3 units.

From the origin, point D is down 3 units.

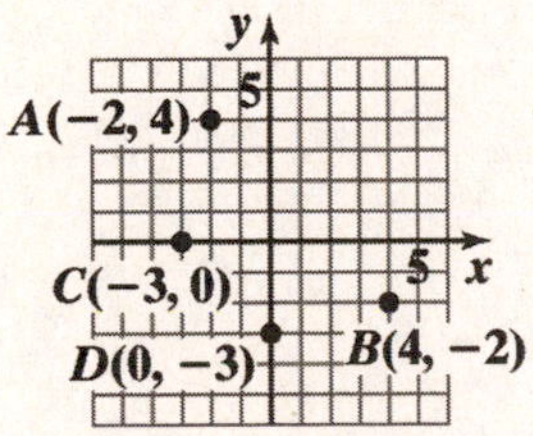

y
5
A(−2, 4)
C(−3, 0)
5 x
D(0, −3)
B(4, −2)

1b. If a point is on the x-axis it is neither up nor down, so $x = 0$.

False. The y-coordinate gives the distance up or down, so $y = 0$.

Pencil Problem #1

1a. Plot the points:
$A(1, 4)$, $B(-2, 3)$, $C(-3, -5)$, and $D(-4, 0)$.

1b. True or false: If a point is on the y-axis, its x-coordinate must be 0.

Objective #2: Graph equations in the rectangular coordinate system.

✔ *Solved Problem #2*

2a. Graph $y = 4 - x$.

x	$y = 4 - x$	(x, y)
−3	$y = 4 - (-3) = 7$	$(-3, 7)$
−2	$y = 4 - (-2) = 6$	$(-2, 6)$
−1	$y = 4 - (-1) = 5$	$(-1, 5)$
0	$y = 4 - (0) = 4$	$(0, 4)$
1	$y = 4 - (1) = 3$	$(1, 3)$
2	$y = 4 - (2) = 2$	$(2, 2)$
3	$y = 4 - (3) = 1$	$(3, 1)$

$y = 4 - x$

2b. Graph $y = |x + 1|$.

x	$y = \|x+1\|$	(x, y)
−4	$y = \|-4+1\| = \|-3\| = 3$	$(-4, 3)$
−3	$y = \|-3+1\| = \|-2\| = 2$	$(-3, 2)$
−2	$y = \|-2+1\| = \|-1\| = 1$	$(-2, 1)$
−1	$y = \|-1+1\| = \|0\| = 0$	$(-1, 0)$
0	$y = \|0+1\| = \|1\| = 1$	$(0, 1)$
1	$y = \|1+1\| = \|2\| = 2$	$(1, 2)$
2	$y = \|2+1\| = \|3\| = 3$	$(2, 3)$

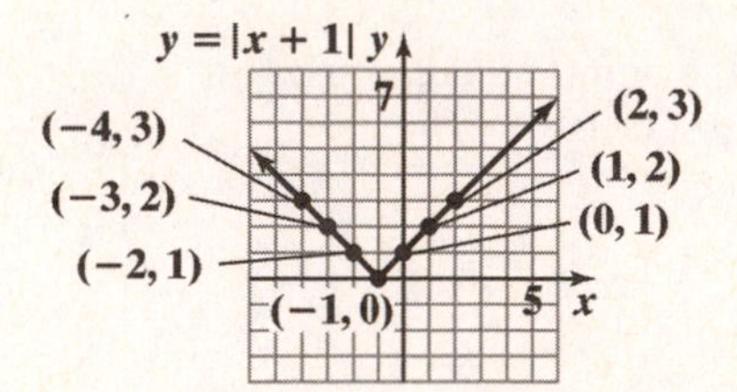

Pencil Problem #2

2a. Graph $y = x^2 - 2$. Let $x = -3, -2, -1, 0, 1, 2,$ and 3.

2b. Graph $y = 2|x|$. Let $x = -3, -2, -1, 0, 1, 2,$ and 3.

Objective #3: Interpret information about a graphing utility's viewing rectangle or table.

✔ *Solved Problem #3*

3. What is the meaning of a
[–100, 100, 50] by [–100, 100, 10]
viewing rectangle?

The minimum x-value is –100, the maximum x-value is 100, and the distance between consecutive tick marks is 50.

The minimum y-value is –100, the maximum y-value is 100, and the distance between consecutive tick marks is 10.

Pencil Problem #3

3. What is the meaning of a
[–20, 80, 10] by [–30, 70, 10]
viewing rectangle?

Objective #4: Use a graph to identify intercepts.

✔ *Solved Problem #4*

4a. Identify the x- and y- intercepts:

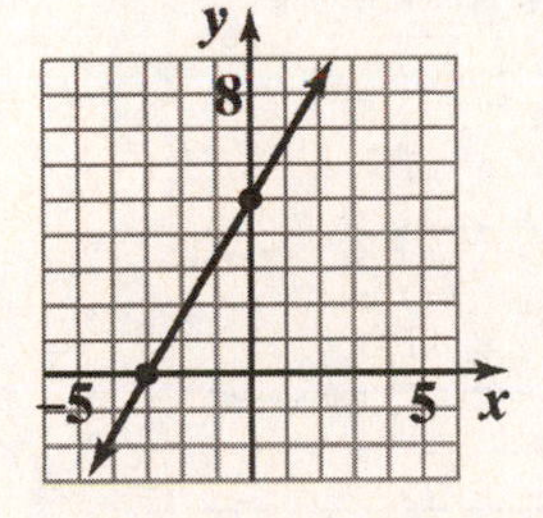

The graph crosses the x-axis at (–3,0).
Thus, the x-intercept is –3.

The graph crosses the y-axis at (0,5).
Thus, the y-intercept is 5.

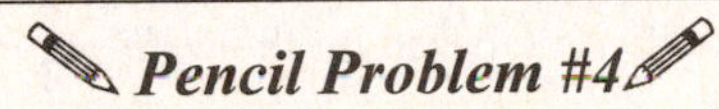

Pencil Problem #4

4a. Identify the x- and y- intercepts:

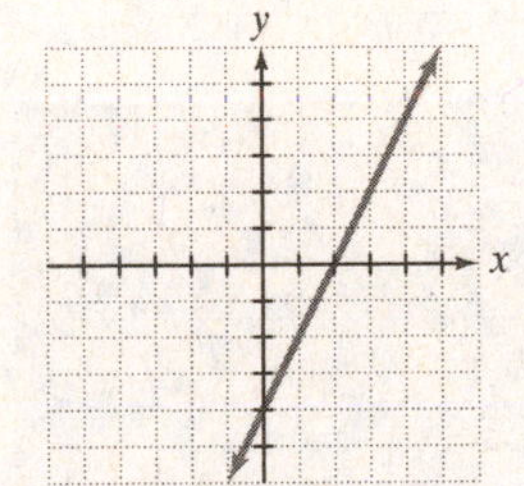

4b. Identify the x- and y- intercepts:

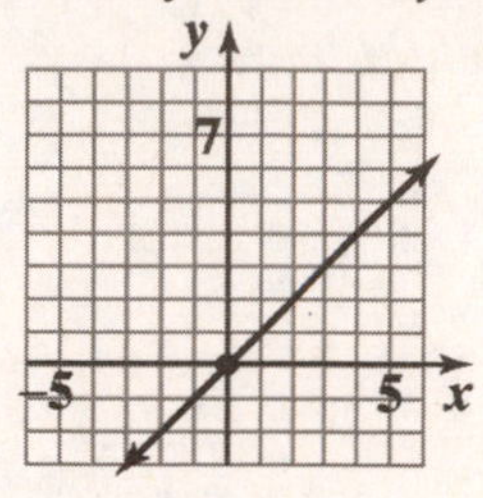

The graph crosses the x-axis at (0,0).
Thus, the x-intercept is 0.

The graph crosses the y-axis at (0,0).
Thus, the y-intercept is 0.

4b. Identify the x- and y- intercepts:

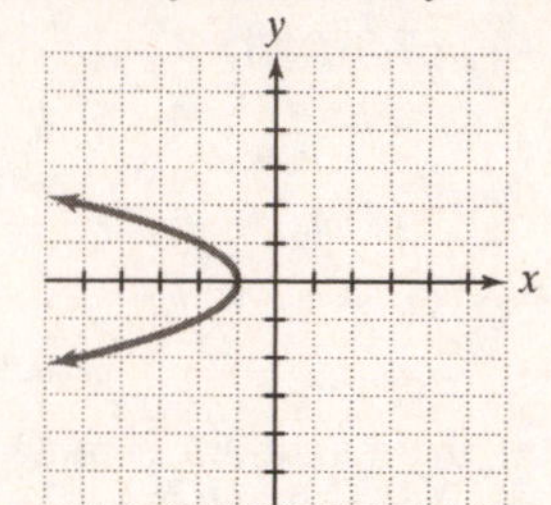

Objective #5: Interpret information given by graphs.

✔ *Solved Problem #5*

5. The line graphs show the percentage of marriages ending in divorce based on the wife's age at marriage.

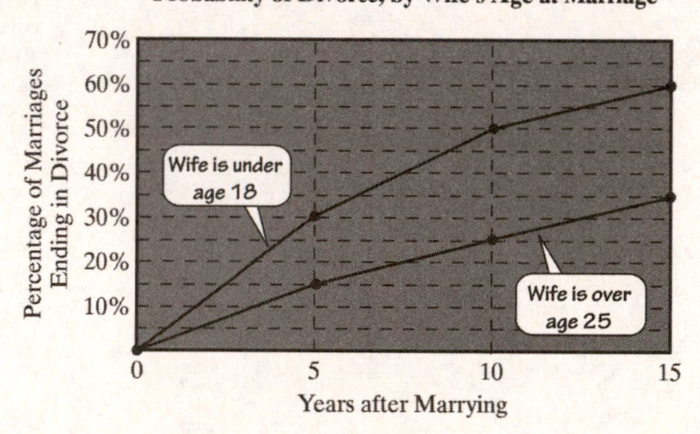

The model $d = 4n + 5$ approximates the data in the graph when the wife is under 18 at the time of marriage. In the model, n is the number of years after marriage and d is the percentage of marriages ending in divorce.

(Continued on next page)

Pencil Problem #5

5. The graphs show the percentage of high school seniors who used alcohol or marijuana.

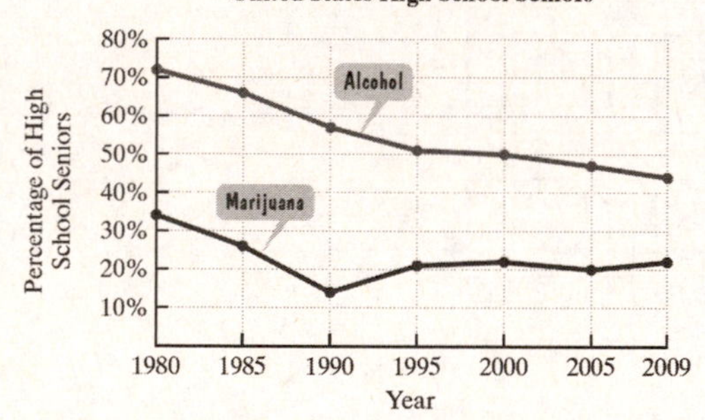

Source: U.S. Department of Health and Human Services

The data for seniors who used marijuana can be modeled by $M = -0.3n + 27$, where M is the percentage of seniors who used marijuana n years after 1980.

(Continued on next page)

5a. Use the formula to determine the percentage of marriages ending in divorce after 15 years when the wife is under 18 at the time of marriage.

$d = 4n + 5$

$d = 4(15) + 5$

$d = 60 + 5$

$d = 65$

According to the formula, 65% of marriages end in divorce after 15 years when the wife is under 18 at the time of marriage.

5a. Use the formula to determine the percentage of seniors who used marijuana in 2005.

5b. Use the appropriate line graph to determine the percentage of marriages ending in divorce after 15 years when the wife is under 18 at the time of marriage.

Locate 15 on the horizontal axis and locate the point above it on the graph. Read across to the corresponding percentage on the vertical axis. This percentage is 60. According to the line graph, 60% of marriages end in divorce after 15 years when the wife is under 18 at the time of marriage.

5b. Use the appropriate line graph to determine the percentage of seniors who used marijuana in 2005.

5c. Does the value given by the model underestimate or overestimate the value shown by the graph? By how much?

The value given by the model, 65%, is greater than the value shown by the graph, 60%, so the model overestimates the percentage by 65 – 5, or 5.

5c. Does the formula underestimate or overestimate the percentage of seniors who used marijuana in 2005 as shown by the graph.

Answers for Pencil Problems *(Textbook Exercise references in parentheses)*:

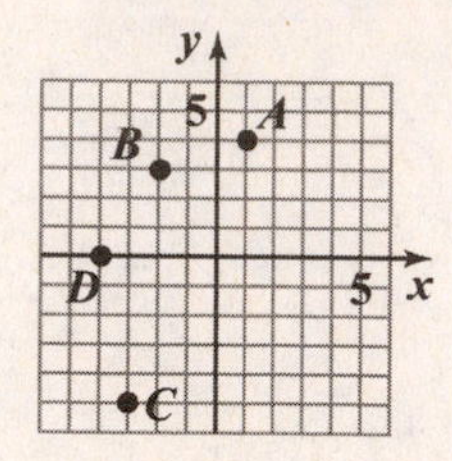

1a. *(1.1 #1-9)*　　**1b.** True *(1.1.#73)*

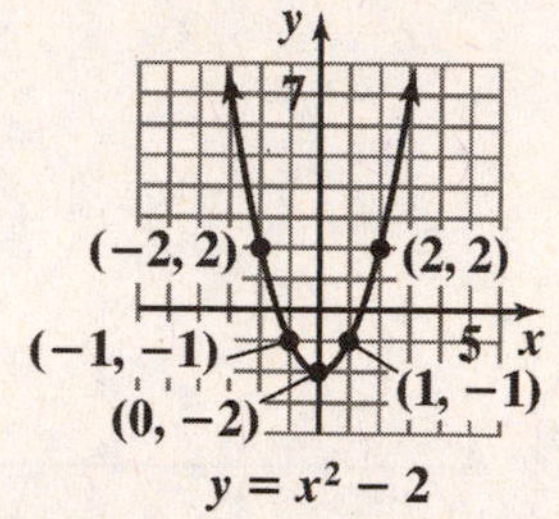

2a. *(1.1 #13)*

$y = 2|x|$ y
7
(−3, 6)
(3, 6)
(−2, 4)
(2, 4)
(−1, 2)
(1, 2)
(0, 0)
5
x

2b. *(1.1 #21)*

3. The minimum x-value is –20, the maximum x-value is 80, and the distance between consecutive tick marks is 10. The minimum y-value is –30, the maximum y-value is 70, and the distance between consecutive tick marks is 10. *(1.1 #31)*

4a. x-intercept: 2; y-intercept: –4 *(1.1 #41)*　　**4b.** x-intercept: –1 ; y-intercept: none *(1.1 #45)*

5a. 19.5% *(1.1 #55b)*　　**5b.** 20% *(1.1 #55a)*　　**5c.** underestimates by 0.5 *(1.1 #55b)*

Section 1.2
Linear Equations and Rational Equations

Up, Up, and Away!

Inflation!
It seems that everything costs more and more each year.

What cost \$10,000 in 1967 would have cost you \$24,900 in 1980 and \$64,500 in 2008!

In the Exercise Set of this section of the textbook, we will look at mathematical formulas that model this increase.

Objective #1: Solve linear equations in one variable

✔ *Solved Problem #1*

1a. Solve and check: $4x+5=29$

$$4x+5=29$$
$$4x+5-5=29-5$$
$$4x=24$$
$$\frac{4x}{4}=\frac{24}{4}$$
$$x=6$$

Check:

$$4x+5=29$$
$$4(6)+5=29$$
$$24+5=29$$
$$29=29$$

The check verifies that the solution set is $\{6\}$.

1b. Solve and check: $4(2x+1)=29+3(2x-5)$

Simplify the algebraic expression on each side.

$$4(2x+1)=29+3(2x-5)$$
$$8x+4=29+6x-15$$
$$8x+4=6x+14$$

Collect variable terms on one side and constant terms on the other side.

$$8x-6x+4=6x-6x+14$$
$$2x+4=14$$
$$2x+4=14$$
$$2x+4-4=14-4$$

(Continued on next page)

Pencil Problem #1

1a. Solve and check: $6x-3=63$

1b. Solve and check: $16=3(x-1)-(x-7)$

Isolate the variable and solve.

$$\frac{2x}{2}=\frac{10}{2}$$
$$x=5$$

Check:

$$4(2x+1)=29+3(2x-5)$$
$$4(2\cdot5+1)=29+3(2\cdot5-5)$$
$$4(11)=29+3(5)$$
$$44=44$$

The solution set is $\{5\}$.

Objective #2: Solve linear equations containing fractions.

✔ *Solved Problem #2*

2. Solve and check: $\frac{x-3}{4}=\frac{5}{14}-\frac{x+5}{7}$

The LCD is 28.

$$\frac{x-3}{4}=\frac{5}{14}-\frac{x+5}{7}$$
$$28\left(\frac{x-3}{4}\right)=28\left(\frac{5}{14}-\frac{x+5}{7}\right)$$
$$\frac{28}{1}\left(\frac{x-3}{4}\right)=\frac{28}{1}\left(\frac{5}{14}\right)-\frac{28}{1}\left(\frac{x+5}{7}\right)$$
$$7(x-3)=2(5)-4(x+5)$$
$$7x-21=10-4x-20$$
$$7x-21=-4x-10$$
$$7x+4x-21=-4x+4x-10$$
$$11x-21=-10$$
$$11x-21+21=-10+21$$
$$11x=11$$
$$\frac{11x}{11}=\frac{11}{11}$$
$$x=1$$

Check:

$$\frac{x-3}{4}=\frac{5}{14}-\frac{x+5}{7}$$
$$\frac{1-3}{4}=\frac{5}{14}-\frac{1+5}{7}$$
$$\frac{-2}{4}=\frac{5}{14}-\frac{6}{7}$$
$$-\frac{1}{2}=-\frac{1}{2}$$

The solution set is $\{1\}$.

✎ *Pencil Problem #2* ✎

2. Solve and check: $\frac{x+3}{6}=\frac{3}{8}+\frac{x-5}{4}$

Objective #3: Solve rational equations with variables in denominators.

✔ *Solved Problem #3*

3a. Solve: $\frac{5}{2x}=\frac{17}{18}-\frac{1}{3x}$

The LCD is $18x$. Two of the denominators would equal 0 if $x=0$, so $x \neq 0$.

$$\frac{5}{2x}=\frac{17}{18}-\frac{1}{3x},\ x\neq 0$$
$$18x\cdot\frac{5}{2x}=18x\cdot\left(\frac{17}{18}-\frac{1}{3x}\right)$$
$$18x\cdot\frac{5}{2x}=18x\cdot\frac{17}{18}-18x\cdot\frac{1}{3x}$$
$$9\cdot 5=x\cdot 17-6\cdot 1$$
$$45=17x-6$$
$$45+6=17x-6+6$$
$$51=17x$$
$$\frac{51}{17}=\frac{17x}{17}$$
$$3=x$$

Note that 3 is not part of the restriction $x \neq 0$. The solution set is $\{3\}$.

✎ *Pencil Problem #3* ✎

3a. Solve: $\frac{4}{x}=\frac{5}{2x}+3$

3b. Solve: $\frac{x}{x-2}=\frac{3}{x-2}-\frac{2}{3}$

The LCD is $3(x-2)$. Two of the denominators would equal 0 if $x=2$, so $x \neq 2$.

$$\frac{x}{x-2}=\frac{2}{x-2}-\frac{2}{3},\ x\neq 2$$
$$3(x-2)\cdot\frac{x}{x-2}=3(x-2)\cdot\left(\frac{2}{x-2}-\frac{2}{3}\right)$$
$$3(x-2)\cdot\frac{x}{x-2}=3(x-2)\cdot\frac{2}{x-2}-3(x-2)\cdot\frac{2}{3}$$
$$3x=3\cdot 2-(x-2)\cdot 2$$
$$3x=6-2x+4$$
$$3x=10-2x$$
$$3x+2x=10-2x+2x$$
$$5x=10$$
$$\frac{5x}{5}=\frac{10}{5}$$
$$x=2$$

The proposed solution is not a solution because of the restriction $x \neq 2$. The solution set is $\varnothing$.

3b. Solve: $\frac{8x}{x+1}=4-\frac{8}{x+1}$

Objective #4: Recognize identities, conditional equations, and inconsistent equations.

✔ Solved Problem #4

4a. Solve and determine whether the equation is an identity, a conditional equation, or an inconsistent equation.
$4x-7=4(x-1)+3$

$$4x-7=4(x-1)+3$$
$$4x-7=4x-4+3$$
$$4x-7=4x-1$$
$$-7=-1$$

This equation is an inconsistent equation and thus has no solution.

The solution set is { } or $\varnothing$.

4b. Solve and determine whether the equation is an identity, a conditional equation, or an inconsistent equation.
$7x+9=9(x+1)-2x$

$$7x+9=9(x+1)-2x$$
$$7x+9=9x+9-2x$$
$$7x+9=7x+9$$
$$9=9$$

This equation is an identity and all real numbers are solutions.

The solution set is $\{x \mid x \text{ is a real number}\}$ or $(-\infty,\infty)$ or $\mathbb{R}$.

✎ Pencil Problem #4 ✎

4a. Solve and determine whether the equation is an identity, a conditional equation, or an inconsistent equation.
$5x+9=9(x+1)-4x$

4b. Solve and determine whether the equation is an identity, a conditional equation, or an inconsistent equation.
$10x+3=8x+3$

Objective #5: Solve applied problems using formulas.

✔ *Solved Problem #5*

5. It has been shown that persons with a low sense of humor have higher levels of depression in response to negative life events than those with a high sense of humor. This can be modeled by the following formulas:

Low-Humor Group: $D=\frac{10}{9}x+\frac{53}{9}$

High-Humor Group: $D=\frac{1}{9}x+\frac{26}{9}$

where x represents the intensity of a negative life event (from a low of 1 to a high of 10) and D is the level of depression in response to that event.

If the low-humor group averages a level of depression of 10 in response to a negative life event, what is the intensity of that event?

Low-Humor Group: $D=\frac{10}{9}x+\frac{53}{9}$

$$10=\frac{10}{9}x+\frac{53}{9}$$

$$9\cdot 10=9\left(\frac{10}{9}x+\frac{53}{9}\right)$$

$$90=10x+53$$

$$90-53=10x+53-53$$

$$37=10x$$

$$\frac{37}{10}=\frac{10x}{10}$$

$$3.7=x$$

$$x=3.7$$

The formula indicates that if the low-humor group averages a level of depression of 10 in response to a negative life event, the intensity of that event is 3.7.

Pencil Problem #5

5. The formula $C=1388x+24,963$ can be used to model the cost, C, x years after 1980 of what cost \$10,000 in 1967. Use the model to determine in which year the cost will be \$77,707 for what cost \$10,000 in 1967.

Answers for Pencil Problems *(Textbook Exercise references in parentheses)*:

1a. {11} *(1.2 #2)* **1b.** {6} *(1.2 #13)* **2.** $\left\{\frac{33}{2}\right\}$ *(1.2 #25)*

3a. $\left\{\frac{1}{2}\right\}$ *(1.2 #31)* **3b.** $\varnothing$ *(1.2 #41)*

4a. The solution set is $\{x \mid x \text{ is a real number}\}$ or $(-\infty, \infty)$ or $\mathbb{R}$. The equation is an identity. *(1.2 #61)*

4b. The solution set is $\{0\}$. The equation is conditional. *(1.2 #65)*

5. 2018 *(1.2 #101)*

Section 1.3
Models and Applications

Counting Your Money!

From how much you can expect to earn at your first job after college to how much you need to save each month for retirement, mathematical models can help you plan your finances. In this section, you will see applications that involve starting salaries with a college degree based on major, investing money in two or more accounts to obtain a specified return each year, and the cost of a text messaging plan.

Objective #1: Use linear equations to solve problems.

✔ *Solved Problem #1*

1a. The median starting salary of a computer science major exceeds that of an education major by $21 thousand. The median starting salary of an economics major exceeds that of an education major by $14 thousand. Combined their median starting salaries are $140 thousand. Determine the median starting salary for each of these three majors.

Since the salaries of both the computer science and economics majors are compared to education majors, we let x = the median starting salary for an education major. The other two majors have salaries that exceed this salary by a specified amount, so we add that amount to the salary for the education major.

$x + 21$ = the median starting salary for a computer science major

$x + 14$ = the median starting salary for an economics major

Since the combined salary is $140 thousand, we add the three salaries and set the sum equal to 140. Then we solve for x.

$$
\begin{aligned}
x+(x+21)+(x+14) &= 140 \\
3x+35 &= 140 \\
3x &= 105 \\
x &= 35
\end{aligned}
$$

$x + 21 = 35 + 21 = 56$
$x + 14 = 35 + 14 = 49$

The median starting salaries are $35 thousand for an education major, $56 thousand for a computer science major, and $49 thousand for an economics major.

You should verify that these salaries are $140 thousand combined.

Pencil Problem #1

1a. According to the American Bureau of Labor Statistics, you will devote 37 years to sleeping and watching TV. The number of years sleeping will exceed the number of years watching TV by 19. Over your lifetime, how many years will you spend on each of these activities?

1b. You are choosing between two texting plans. Plan A has a monthly fee of $15 with a charge of $0.08 per text. Plan B has a monthly fee of $3 with a charge of $0.12 per text. For how many text messages will the cost of the two plans be the same?

Let x = the number of text messages for which the cost of the two plans is the same.

The monthly cost for Plan A is $15 plus $0.08 for each text times the number of texts, x:

$15 + 0.08x$

The monthly cost for Plan B is $3 plus $0.12 for each text times the number of texts, x:

$3 + 0.12x$

Since we are interested in the costs being the same, we set the costs equal and solve for x.

$$\begin{aligned}15+0.08x &= 3+0.12x\\ 15 &= 3+0.04x\\ 12 &= 0.04x\\ \frac{12}{0.04} &= \frac{0.04x}{0.04}\\ 300 &= x\end{aligned}$$

The costs are the same for 300 text messages.

You should verify that both plans cost $39 when 300 text messages are used.

1b. You are choosing between two health clubs. Club A offers membership for a fee of $40 plus a monthly fee of $25. Club B offers membership for a fee of $15 plus a monthly fee of $30. After how many months will the total cost at each health club be the same?

1c. You inherited \$5000 with the stipulation that for the first year the money had to be invested in two funds paying 9% and 11% annual interest. How much did you invest at each rate if the total interest earned for the year was \$487?

Let x = the amount invested at 9%.

Since a total of \$5000 is invested, $5000 - x$ is invested at 11%.

Note that $x + (5000 - x) = 5000$.

The interest on the amount invested at 9% is $0.09x$, using $I = Pr$. The interest on the amount invested at 11% is $0.11(5000 - x)$. The total interest is 487.

$$\begin{aligned} 0.09x + 0.11(5000 - x) &= 487 \\ 0.09x + 550 - 0.11x &= 487 \\ -0.02x + 550 &= 487 \\ -0.02x &= -63 \\ \frac{-0.02x}{-0.02} &= \frac{-63}{-0.02} \\ x &= 3150 \end{aligned}$$

$5000 - x = 5000 - 3150 = 1850$

\$3150 was invested at 9%, and \$1850 was invested at 11%.

You should verify that the resulting interest is \$487.

1c. You invested \$7000 in two accounts paying 6% and 8% annual interest. If the total interest earned for the year was \$520, how much was invested at each rate?

Objective #2: Solve a formula for a variable.

Solved Problem #2

2. Solve each formula for the specified variable.

2a. $P = 2l + 2w$ for w

$$P = 2l + 2w$$
$$P - 2l = 2l - 2l + 2w$$
$$P - 2l = 2w$$
$$\frac{P-2l}{2} = \frac{2w}{2}$$
$$\frac{P-2l}{2} = w \text{ or } w = \frac{P-2l}{2}$$

2b. $P = C + MC$ for C

Begin by factoring out C on the right.

$$P = C + MC$$
$$P = C(1+M)$$
$$\frac{P}{1+M} = \frac{C(1+M)}{1+M}$$
$$\frac{P}{1+M} = C \text{ or } C = \frac{P}{1+M}$$

Pencil Problem #2

2. Solve each formula for the specified variable.

2a. $T = D + pm$ for p

2b. $IR + Ir = E$ for I

Answers for Pencil Problems *(Textbook Exercise references in parentheses)*:

1a. TV: 9 years; sleeping: 28 years *(1.3 #19)*

1b. 5 months *(1.3 #27)*

1c. $2000 at 6% and $5000 at 8% *(1.3 #39)*

2a. $p = \frac{T-D}{m}$ *(1.3 #63)* **2b.** $I = \frac{E}{R+r}$ *(1.3 #71)*

Section 1.4
Complex Numbers

Why Study Something if it is *IMAGINARY*???

Great Question!
The numbers that we study in this section were given the name "*imaginary*" at a time when mathematicians believed such numbers to be useless.

Since that time, many *real-life* applications for so-called imaginary numbers have been discovered, but the name they were originally given has endured.

Objective #1: Add and subtract complex numbers.

✔ *Solved Problem #1*

1a. Add: $(5-2i)+(3+3i)$

$$(5-2i)+(3+3i)=5-2i+3+3i$$
$$=8+i$$

1b. Subtract: $(2+6i)-(12-4i)$

$$(2+6i)-(12-4i)=2+6i-12+4i$$
$$=-10+10i$$

Pencil Problem #1

1a. Add: $(7+2i)+(1-4i)$

1b. Subtract: $(3+2i)-(5-7i)$

Objective #2: Multiply complex numbers.

✔ *Solved Problem #2*

2a. Multiply: $(5+4i)(6-7i)$

$$\begin{aligned}(5+4i)(6-7i) &= 30-35i+24i-28i^2\\ &= 30-35i+24i-28(-1)\\ &= 30+28-35i+24i\\ &= 58-11i\end{aligned}$$

2b. Multiply: $7i(2-9i)$

$$\begin{aligned}7i(2-9i) &= 7i\cdot 2-7i\cdot 9i\\ &= 14i-63i^2\\ &= 14i-63(-1)\\ &= 63+14i\end{aligned}$$

Pencil Problem #2

2a. Multiply: $(-5+4i)(3+i)$

2b. Multiply: $-3i(7i-5)$

Objective #3: Divide complex numbers.

✔ *Solved Problem #3*

3. Divide and express the result in standard form:
$\dfrac{5+4i}{4-i}$

Multiply the numerator and the denominator by the conjugate of the denominator, $4+i$.

$$\begin{aligned}\frac{5+4i}{4-i} &= \frac{(5+4i)}{(4-i)}\cdot\frac{(4+i)}{(4+i)}\\ &= \frac{20+5i+16i+4i^2}{16+1}\\ &= \frac{20+21i+4(-1)}{16+1}\\ &= \frac{16+21i}{17}\\ &= \frac{16}{17}+\frac{21}{17}i\end{aligned}$$

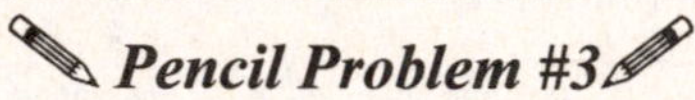

Pencil Problem #3

3. Divide and express the result in standard form:
$\dfrac{2+3i}{2+i}$

Objective #4: Perform the indicated operations and write the result in standard form.

✔ Solved Problem #4

4. Perform the indicated operations and write the result in standard form.

4a. $\sqrt{-27}+\sqrt{-48}$

$$\begin{aligned}\sqrt{-27}+\sqrt{-48} &= i\sqrt{27}+i\sqrt{48}\\ &= i\sqrt{9\cdot 3}+i\sqrt{16\cdot 3}\\ &= 3i\sqrt{3}+4i\sqrt{3}\\ &= 7i\sqrt{3}\end{aligned}$$

4b. $(-2+\sqrt{-3})^2$

$$\begin{aligned}(-2+\sqrt{-3})^2 &= (-2+i\sqrt{3})^2\\ &= (-2)^2+2(-2)(i\sqrt{3})+(i\sqrt{3})^2\\ &= 4-4i\sqrt{3}+3i^2\\ &= 4-4i\sqrt{3}+3(-1)\\ &= 1-4i\sqrt{3}\end{aligned}$$

4c. $\dfrac{-14+\sqrt{-12}}{2}$

$$\begin{aligned}\frac{-14+\sqrt{-12}}{2} &= \frac{-14+i\sqrt{12}}{2}\\ &= \frac{-14+2i\sqrt{3}}{2}\\ &= \frac{-14}{2}+\frac{2i\sqrt{3}}{2}\\ &= -7+i\sqrt{3}\end{aligned}$$

✎ Pencil Problem #4 ✎

4. Perform the indicated operations and write the result in standard form.

4a. $\sqrt{-64}-\sqrt{-25}$

4b. $(-3-\sqrt{-7})^2$

4c. $\dfrac{-8+\sqrt{-32}}{24}$

Answers for Pencil Problems *(Textbook Exercise references in parentheses)*:

1a. $8-2i$ *(1.4 #1)* **1b.** $-2+9i$ *(1.4 #3)*

2a. $-19+7i$ *(1.4 #11)* **2b.** $21+15i$ *(1.4 #9)*

3. $\frac{7}{5}+\frac{4}{5}i$ *(1.4 #27)*

4a. $3i$ *(1.4 #29)* **4b.** $2+6i\sqrt{7}$ *(1.4 #35)* **4c.** $-\frac{1}{3}+i\frac{\sqrt{2}}{6}$ *(1.4 #37)*

Section 1.5
Quadratic Equations

Maybe I Should Ride the Bus Instead

Did you know that the likelihood that a driver will be involved in a fatal crash decreases with age until about age 45 and then increases after that? Formulas that model data that first decrease and then increase contain a variable squared. When we use these models to answer questions about the data, we often need to find the solutions of a *quadratic equation*.

Unlike linear equations, quadratic equations may have exactly two distinct solutions. Thus, when we find the age at which drivers are involved in 3 fatal crashes per 100 million miles driven, we will find two different ages, one less 45 and the other greater than 45.

Objective #1: Solve quadratic equations by factoring.

✔ *Solved Problem #1*

1. Solve by factoring.

1a. $3x^2 - 9x = 0$

$$\begin{aligned} 3x^2 - 9x &= 0 \\ 3x(x-3) &= 0 \\ 3x = 0 \text{ or } x - 3 &= 0 \\ x = 0 \text{ or } \quad x &= 3 \end{aligned}$$

The solution set is $\{0, 3\}$.

1b. $2x^2 + x = 1$

$$\begin{aligned} 2x^2 + x &= 1 \\ 2x^2 + x - 1 &= 0 \\ (2x-1)(x+1) &= 0 \\ 2x - 1 = 0 \text{ or } x + 1 &= 0 \\ 2x = 1 \text{ or } \quad x &= -1 \\ x &= \frac{1}{2} \end{aligned}$$

The solution set is $\left\{-1, \frac{1}{2}\right\}$.

Pencil Problem #1

1. Solve by factoring.

1a. $3x^2 + 12x = 0$

1b. $x^2 = 8x - 15$

Objective #2: Solve quadratic equations by the square root property.

✔ *Solved Problem #2*

2. Solve by the square root property.

2a. $3x^2 - 21 = 0$

$$3x^2 - 21 = 0$$
$$3x^2 = 21$$
$$x^2 = 7$$
$$x = \pm\sqrt{7}$$

The solution set is $\left\{-\sqrt{7}, \sqrt{7}\right\}$.

2b. $5x^2 + 45 = 0$

$$5x^2 + 45 = 0$$
$$5x^2 = -45$$
$$x^2 = -9$$
$$x = \pm\sqrt{-9}$$
$$x = \pm 3i$$

The solution set is $\{-3i, 3i\}$.

2c. $(x+5)^2 = 11$

$$(x+5)^2 = 11$$
$$x + 5 = \pm\sqrt{11}$$
$$x = -5 \pm \sqrt{11}$$

The solution set is $\{-5+\sqrt{11}, -5-\sqrt{11}\}$.

Pencil Problem #2

2. Solve by the square root property.

2a. $5x^2 + 1 = 51$

2b. $2x^2 - 5 = -55$

2c. $3(x-4)^2 = 15$

Objective #3: Solve quadratic equations by completing the square.

✔ *Solved Problem #3*

3a. What term should be added to the binomial $x^2 + 6x$ so that it becomes a perfect square trinomial? Write and factor the trinomial.

The coefficient of the x-term of $x^2 + 6x$ is 6.

Half of 6 is 3, and 3^2 is 9, which should be added to the binomial.

The result is a perfect square trinomial.

$$x^2 + 6x + 9 = (x+3)^2$$

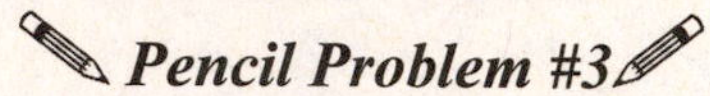

Pencil Problem #3

3a. What term should be added to the binomial $x^2 - 10x$ so that it becomes a perfect square trinomial? Write and factor the trinomial.

3b. Solve by completing the square: $x^2+4x-1=0$

$$x^2+4x-1=0$$
$$x^2+4x \quad =1$$

Half of 4 is 2, and 2^2 is 4, which should be added to both sides.

$$x^2+4x+4=1+4$$
$$x^2+4x+4=5$$
$$(x+2)^2=5$$
$$x+2=\sqrt{5} \quad \text{or} \quad x+2=-\sqrt{5}$$
$$x=-2+\sqrt{5} \qquad x=-2-\sqrt{5}$$

The solution set is $\left\{-2\pm\sqrt{5}\right\}$.

3b. Solve by completing the square: $x^2+6x-11=0$

3c. Solve by completing the square: $2x^2+3x-4=0$

Since the coefficient of x^2 is 2, begin by dividing both sides by 2.

$$2x^2+3x-4=0$$
$$x^2+\frac{3}{2}x-2=0$$
$$x^2+\frac{3}{2}x=2$$

Half of the coefficient of x is $\frac{1}{2}\left(\frac{3}{2}\right)=\frac{3}{4}$, and $\left(\frac{3}{4}\right)^2=\frac{9}{16}$. Add $\frac{9}{16}$ to both sides.

$$x^2+\frac{3}{2}x+\frac{9}{16}=2+\frac{9}{16}$$
$$\left(x+\frac{3}{4}\right)^2=\frac{41}{16}$$
$$x+\frac{3}{4}=\pm\sqrt{\frac{41}{16}}$$
$$x=-\frac{3}{4}\pm\frac{\sqrt{41}}{4}$$
$$x=\frac{-3\pm\sqrt{41}}{4}$$

The solution set is $\left\{\frac{-3+\sqrt{41}}{4}, \frac{-3-\sqrt{41}}{4}\right\}$.

3c. Solve by completing the square: $3x^2-2x-2=0$

Objective #4: Solve quadratic equations using the quadratic formula.

Solved Problem #4

4a. Solve using the quadratic formula: $2x^2+2x-1=0$

The equation is in the form $ax^2+bx+c=0$, where $a=2$, $b=2$, and $c=-1$.

$$x=\frac{-b\pm\sqrt{b^2-4ac}}{2a}$$
$$=\frac{-2\pm\sqrt{2^2-4(2)(-1)}}{2(2)}$$
$$=\frac{-2\pm\sqrt{4+8}}{4}$$
$$=\frac{-2\pm\sqrt{12}}{4}$$
$$=\frac{-2\pm2\sqrt{3}}{4}$$
$$=\frac{2(-1\pm\sqrt{3})}{4}$$
$$=\frac{-1\pm\sqrt{3}}{2}$$

The solution set is $\left\{\frac{-1+\sqrt{3}}{2}, \frac{-1-\sqrt{3}}{2}\right\}$.

4b. Solve using the quadratic formula: $x^2-2x+2=0$

The equation is in the form $ax^2+bx+c=0$, where $a=1$, $b=-2$, and $c=2$.

$$x=\frac{-b\pm\sqrt{b^2-4ac}}{2a}$$
$$=\frac{-(-2)\pm\sqrt{(-2)^2-4(1)(2)}}{2(1)}$$
$$=\frac{2\pm\sqrt{4-8}}{2}$$
$$=\frac{2\pm\sqrt{-4}}{2}$$
$$=\frac{2\pm2i}{2}$$
$$=\frac{2(1\pm i)}{2}$$
$$=1\pm i$$

The solution set is $\{1+i, 1-i\}$.

Pencil Problem #4

4a. Solve using the quadratic formula: $3x^2-3x-4=0$

4b. Solve using the quadratic formula: $x^2-6x+10=0$

Objective #5: Use the discriminant to determine the number and type of solutions.

✔ *Solved Problem #5*

5a. Compute the discriminant and determine the number and type of solutions: $x^2+6x+9=0$

$$b^2-4ac=6^2-4(1)(9)$$
$$=0$$

Since the discriminant is zero, there is one (repeated) real rational solution.

5b. Compute the discriminant and determine the number and type of solutions: $2x^2-7x-4=0$

$$b^2-4ac=(-7)^2-4(2)(-4)$$
$$=81$$

Since the discriminant is positive and a perfect square, there are two real rational solutions.

5c. Compute the discriminant and determine the number and type of solutions: $3x^2-2x+4=0$

$$b^2-4ac=(-2)^2-4(3)(4)$$
$$=-44$$

Since the discriminant is negative, there is no real solution. There are imaginary solutions that are complex conjugates.

Pencil Problem #5

5a. Compute the discriminant and determine the number and type of solutions: $x^2-2x+1=0$

5b. Compute the discriminant and determine the number and type of solutions: $x^2-4x-5=0$

5c. Compute the discriminant and determine the number and type of solutions: $4x^2-2x+3=0$

Objective #6: Determine the most efficient method to use when solving a quadratic equation.

✔ *Solved Problem #6*

6. What is the most efficient method for solving a quadratic equation of the form $ax^2+c=0$?

The most efficient method is to solve for x^2 and apply the square root property.

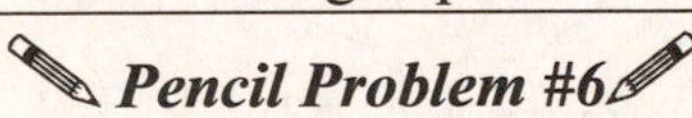

Pencil Problem #6

6. What is the most efficient method for solving a quadratic equation of the form $u^2=d$, where u is a first-degree polynomial?

Objective #7: Solve problems modeled by quadratic equations.

✔ *Solved Problem #7*

7. The function $P(A) = 0.01A^2 + 0.05A + 107$ models a woman's normal systolic blood pressure, $P(A)$, at age A. Use this function to find the age, to the nearest year, of a woman whose normal systolic blood pressure is 115 mm Hg.

$$P(A) = 0.01A^2 + 0.05A + 107$$
$$115 = 0.01A^2 + 0.05A + 107$$
$$0 = 0.01A^2 + 0.05A - 8$$

$a = 0.01 \quad b = 0.05 \quad c = -8$

$$x = \frac{-b \pm \sqrt{b^2 - 4ac}}{2a}$$
$$x = \frac{-0.05 \pm \sqrt{0.05^2 - 4(0.01)(-8)}}{2(0.01)}$$
$$= \frac{-0.05 \pm \sqrt{0.3225}}{0.02}$$
$$x = \frac{-0.05 + \sqrt{0.3225}}{0.02} \quad \text{or} \quad x = \frac{-0.05 - \sqrt{0.3225}}{0.02}$$
$$x \approx 26 \qquad \cancel{x \approx -31}$$

A woman's normal systolic blood pressure is 115 mm at about 26 years of age.

Pencil Problem #7

7. The number of fatal crashes per 100 million miles driven, F, for drivers of age x can be modeled by the formula $F = 0.013x^2 - 1.19x + 28.24$. What age groups are expected to be in 3 fatal crashes per 100 million miles driven.

Answers for Pencil Problems *(Textbook Exercise references in parentheses)*:

1a. $\{-4, 0\}$ *(1.5 #9)* **1b.** $\{3, 5\}$ *(1.5 #3)*

2a. $\{-\sqrt{10}, \sqrt{10}\}$ *(1.5 #17)* **2b.** $\{-5i, 5i\}$ *(1.5 #19)* **2c.** $\{4+\sqrt{5}, 4-\sqrt{5}\}$ *(1.5 #23)*

3a. 25; $x^2 - 10x + 25 = (x-5)^2$ *(1.5 #37)* **3b.** $\{3+2\sqrt{5}, 3-2\sqrt{5}\}$ *(1.5 #51)*

3c. $\left\{\frac{1+\sqrt{7}}{3}, \frac{1-\sqrt{7}}{3}\right\}$ *(1.5 #63)*

4a. $\left\{\frac{3+\sqrt{57}}{6}, \frac{3-\sqrt{57}}{6}\right\}$ *(1.5 #69)* **4b.** $\{3+i, 3-i\}$ *(1.5 #73)*

5a. 0; one (repeated) real rational solution *(1.5 #79)* **5b.** 36; two real rational solutions *(1.5 #75)*

5c. −44; two imaginary solutions that are complex conjugates *(1.5 #76)*

6. The square root property *(1.5 #15-34)*

7. 33-year-olds and 58-year-olds *(1.5 #135)*

Section 1.6
Other Types of Equations

Slam Dunk!

A basketball player's hang time is the time spent in the air when shooting a basket.

In this section, we will be given a formula that involves radicals which models seconds of hang time in terms of the vertical distance of a player's jump.

Objective #1: Solve polynomial equations by factoring.

✔ *Solved Problem #1*

1a. Solve by factoring: $4x^4 = 12x^2$

$$4x^4 = 12x^2$$
$$4x^4 - 12x^2 = 0$$
$$4x^2(x^2 - 3) = 0$$
$$4x^2 = 0 \text{ or } x^2 - 3 = 0$$
$$x^2 = 0 \qquad x^2 = 3$$
$$x = 0 \qquad x = \pm\sqrt{3}$$

The solution set is $\{-\sqrt{3},\ 0,\ \sqrt{3}\}$.

1b. Solve by factoring: $2x^3 + 3x^2 = 8x + 12$

$$2x^3 + 3x^2 = 8x + 12$$
$$2x^3 + 3x^2 - 8x - 12 = 0$$
$$x^2(2x+3) - 4(2x+3) = 0$$
$$(2x+3)(x^2 - 4) = 0$$
$$2x + 3 = 0 \quad \text{or} \quad x^2 - 4 = 0$$
$$2x = -3 \qquad x^2 = 4$$
$$x = -\frac{3}{2} \qquad x = \pm 2$$

The solution set is $\left\{-2,\ -\frac{3}{2},\ 2\right\}$.

Pencil Problem #1

1a. Solve by factoring: $3x^4 - 48x^2 = 0$

1b. Solve by factoring: $3x^3 + 2x^2 = 12x + 8$

Objective #2: Solve radical equations.

✔ *Solved Problem #2*

2a. Solve: $\sqrt{x+3}+3=x$

$$\sqrt{x+3}+3=x$$
$$\sqrt{x+3}=x-3$$
$$(\sqrt{x+3})^2=(x-3)^2$$
$$x+3=x^2-6x+9$$
$$0=x^2-7x+6$$
$$0=(x-6)(x-1)$$
$$x-6=0 \quad \text{or} \quad x-1=0$$
$$x=6 \qquad\qquad x=1$$

Check 6: $\sqrt{x+3}+3=x$

$$\sqrt{6+3}+3=6$$
$$6=6$$

Check 1: $\sqrt{x+3}+3=x$

$$\sqrt{1+3}+3=1$$
$$5=1$$

The solution set is $\{6\}$.

2b. Solve: $\sqrt{x+5}-\sqrt{x-3}=2$

$$\sqrt{x+5}-\sqrt{x-3}=2$$
$$\sqrt{x+5}=\sqrt{x-3}+2$$
$$\left(\sqrt{x+5}\right)^2=\left(\sqrt{x-3}+2\right)^2$$
$$x+5=x-3+4\sqrt{x-3}+4$$
$$x+5=x+1+4\sqrt{x-3}$$
$$4=4\sqrt{x-3}$$
$$1=\sqrt{x-3}$$
$$1^2=\left(\sqrt{x-3}\right)^2$$
$$1=x-3$$
$$4=x$$

Check:

$$\sqrt{x+5}-\sqrt{x-3}=2$$
$$\sqrt{4+5}-\sqrt{4-3}=2$$
$$2=2$$

The solution set is $\{4\}$.

✎ *Pencil Problem #2*

2a. Solve: $\sqrt{2x+13}=x+7$

2b. Solve: $\sqrt{x-5}-\sqrt{x-8}=3$

Objective #3: Solve equations with rational exponents.

✔ Solved Problem #3

3a. Solve: $5x^{\frac{3}{2}} - 25 = 0$

$$5x^{\frac{3}{2}} - 25 = 0$$
$$5x^{\frac{3}{2}} = 25$$
$$x^{\frac{3}{2}} = 5$$
$$(x^{\frac{3}{2}})^{\frac{2}{3}} = (5)^{\frac{2}{3}}$$
$$x = 5^{\frac{2}{3}} \text{ or } \sqrt[3]{25}$$

Check:
$$5x^{\frac{3}{2}} - 25 = 0$$
$$5(5^{\frac{2}{3}})^{\frac{3}{2}} - 25 = 0$$
$$5(5) - 25 = 0$$
$$0 = 0$$

The solution set is $\{5^{\frac{2}{3}}\}$ or $\{\sqrt[3]{25}\}$.

3b. Solve: $x^{\frac{2}{3}} - 8 = -4$

$$x^{\frac{2}{3}} - 8 = -4$$
$$x^{\frac{2}{3}} = 4$$
$$(x^{\frac{2}{3}})^{\frac{3}{2}} = \pm 4^{\frac{3}{2}}$$
$$x = \pm 8$$

You should verify that both −8 and 8 are solutions.
The solution set is $\{-8, 8\}$.

Pencil Problem #3

3a. Solve: $6x^{\frac{5}{2}} - 12 = 0$

3b. Solve: $(x-4)^{\frac{2}{3}} = 16$

Objective #4: Solve equations that are quadratic in form.

✔ *Solved Problem #4*

4a. Solve: $x^4 - 5x^2 + 6 = 0$

Let $u = x^2$.

$$x^4 - 5x^2 + 6 = 0$$
$$\left(x^2\right)^2 - 5x^2 + 6 = 0$$
$$u^2 - 5u + 6 = 0$$
$$(u-3)(u-2) = 0$$

Apply the zero product principle.

$u - 3 = 0$ or $u - 2 = 0$

$u = 3$ $\quad$ $u = 2$

Replace u with x^2.

$x^2 = 3$ or $x^2 = 2$

$x = \pm\sqrt{3}$ $\quad$ $x = \pm\sqrt{2}$

The solution set is $\left\{\pm\sqrt{2}, \pm\sqrt{3}\right\}$.

Pencil Problem #4

4a. Solve: $x^4 - 5x^2 + 4 = 0$

✔ *Solved Problem #4*

4b. Solve: $3x^{\frac{2}{3}} - 11x^{\frac{1}{3}} - 4 = 0$

Rewrite as follows.

$$3(x^{\frac{1}{3}})^2 - 11x^{\frac{1}{3}} - 4 = 0$$

Let $u = x^{\frac{1}{3}}$.

$$3u^2 - 11u - 4 = 0$$
$$(3u+1)(u-4) = 0$$

$3u + 1 = 0$ or $u - 4 = 0$

$3u = -1$ $\quad$ $u = 4$

$u = -\frac{1}{3}$

$x^{\frac{1}{3}} = -\frac{1}{3}$ or $x^{\frac{1}{3}} = 4$

$(x^{\frac{1}{3}})^3 = \left(-\frac{1}{3}\right)^3$ $\quad$ $(x^{\frac{1}{3}})^3 = (4)^3$

$x = -\frac{1}{27}$ $\quad$ $x = 64$

The solution set is $\left\{-\frac{1}{27}, 64\right\}$.

Pencil Problem #4

4b. Solve: $x^{\frac{2}{3}} - x^{\frac{1}{3}} - 6 = 0$

Objective #5: Solve equations involving absolute value.

✔ *Solved Problem #5*

5. Solve: $|2x-1|=5$

Rewrite without absolute value bars.
$|u|=c$ means $u=c$ or $u=-c$.

$$\begin{aligned} 2x-1&=5 &\quad\text{or}\quad 2x-1&=-5\\ 2x&=6 & 2x&=-4\\ x&=3 & x&=-2 \end{aligned}$$

The solution set is $\{-2,3\}$.

✎ *Pencil Problem #5* ✎

5. Solve: $|x-2|=7$

Objective #6: Solve problems modeled by equations.

✔ *Solved Problem #6*

6. The formula $H=-2.3\sqrt{I}+67.6$ models weekly television viewing time, H, in hours, by annual income, I, in thousands of dollars. What annual income corresponds to 33.1 hours per week watching TV?

Substitute 33.1 for H and solve for I.

$$\begin{aligned} 33.1&=-2.3\sqrt{I}+67.6\\ -34.5&=-2.3\sqrt{I}\\ 15&=\sqrt{I}\\ (15)^2&=(\sqrt{I})^2\\ 225&=I \end{aligned}$$

An annual income of $225,000 corresponds to 33.1 hours per week watching TV.

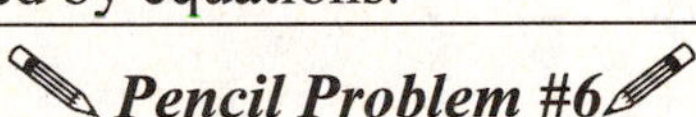

✎ *Pencil Problem #6* ✎

6. The formula $t=\frac{\sqrt{d}}{2}$ models a basketball player's hang time, t, in seconds, in terms of the vertical distance, d, in feet. If the hang time is 1.16 seconds, what is the vertical distance of the jump, to the nearest tenth of a foot?

Answers for Pencil Problems *(Textbook Exercise references in parentheses)*:

1a. $\{-4, 0, 4\}$ *(1.6 #1)* **1b.** $\left\{-2, -\frac{2}{3}, 2\right\}$ *(1.6 #3)*

2a. $\{-6\}$ *(1.6 #15)* **2b.** $\varnothing$ *(1.6 #25)*

3a. $\{\sqrt[5]{4}\}$ *(1.6 #35)* **3b.** $\{-60, 68\}$ *(1.6 #37)*

4a. $\{-2, -1, 1, 2\}$ *(1.6 #41)* **4b.** $\{-8, 27\}$ *(1.6 #49)*

5. $\{-5, 9\}$ *(1.6 #63)*

6. 5.4 ft *(1.6 #105)*

Section 1.7
Linear Inequalities and Absolute Values Inequalities

Are You in LOVE?

As the years go by in a relationship, three key components of love…

passion

commitment

intimacy

…progress differently over time.

Passion peaks early in a relationship and then declines.
By contrast, intimacy and commitment build gradually.

In the applications of this section of the textbook, we will use mathematics to explore the relationships among these three variables of love.

Objective #1: Use interval notation.

✔ *Solved Problem #1*

1a. Express $[-2,5)$ in set-builder notation and graph.

$\{x|-2 \le x < 5\}$

−5 −4 −3 −2 −1 0 1 2 3 4 5

1b. Express $(-\infty,-1)$ in set-builder notation and graph.

$\{x|x < -1\}$

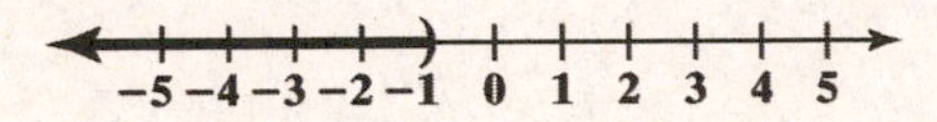

Pencil Problem #1

1a. Express $(1,6]$ in set-builder notation and graph.

1b. Express $[-3,\infty)$ in set-builder notation and graph.

Objective #2: Find intersections and unions of intervals.

✔ *Solved Problem #2*

2. Use graphs to find each set:

2a. $[1, 3] \cap (2, 6)$

Graph each interval. The intersection consists of the portion of the number line that the two graphs have in common.

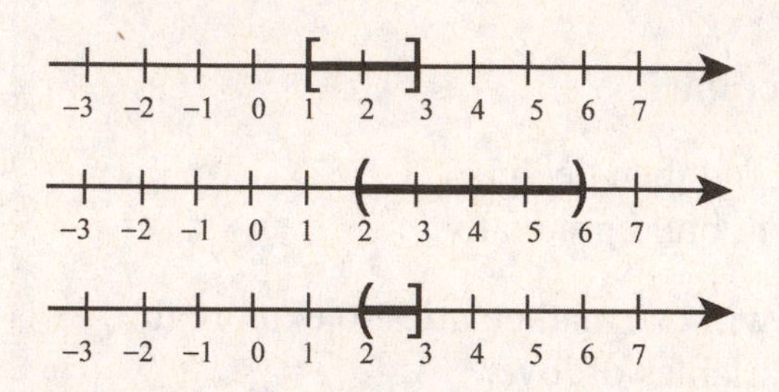

$[1, 3] \cap (2, 6) = (2, 3]$

2b. $[1, 3] \cup (2, 6)$

Graph each interval. The union consists of the portion of the number line in either one of the intervals or the other or both.

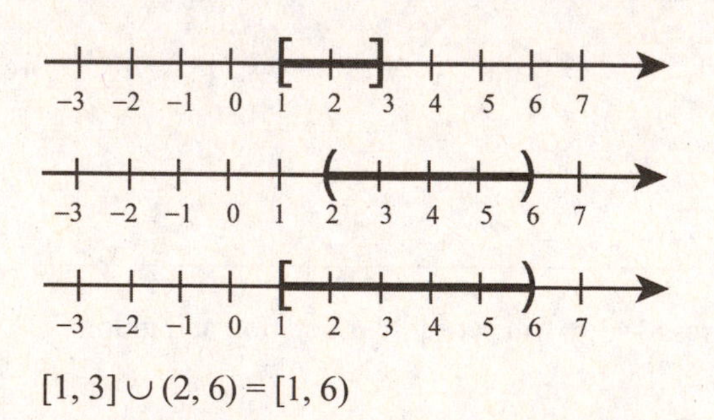

$[1, 3] \cup (2, 6) = [1, 6)$

Pencil Problem #2

2. Use graphs to find each set:

2a. $(-3, 0) \cap [-1, 2]$

2b. $(-3, 0) \cup [-1, 2]$

Objective #3: Solve linear inequalities.

✔ *Solved Problem #3*

3a. Solve and graph the solution set on a number line:

$$3x+1>7x-15$$

$$3x+1>7x-15$$
$$-4x>-16$$
$$\frac{-4x}{-4}<\frac{-16}{-4}$$
$$x<4$$

The solution set is $(-\infty,4)$.

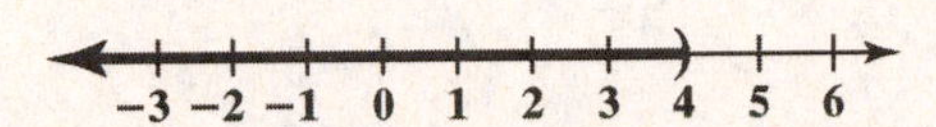

3b. Solve and graph the solution set on a number line:

$$\frac{x-4}{2}\geq\frac{x-2}{3}+\frac{5}{6}$$

$$\frac{x-4}{2}\geq\frac{x-2}{3}+\frac{5}{6}$$
$$6\left(\frac{x-4}{2}\right)\geq 6\left(\frac{x-2}{3}+\frac{5}{6}\right)$$
$$3(x-4)\geq 2(x-2)+5$$
$$3x-12\geq 2x-4+5$$
$$3x-12\geq 2x+1$$
$$x\geq 13$$

The solution set is $[13,\infty)$.

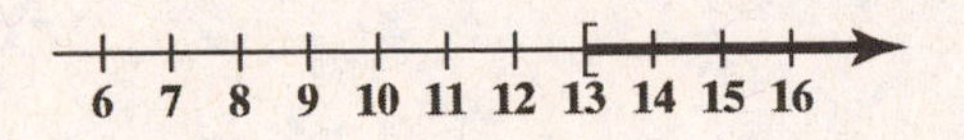

 *Pencil Problem #2*

3a. Solve and graph the solution set on a number line:

$$-9x\geq 36$$

3b. Solve and graph the solution set on a number line:

$$\frac{x}{4}-\frac{3}{2}=\frac{x}{2}+1$$

3c. A car can be rented from Basic Rental for $260 per week with no extra charge for mileage. Continental charges $80 per week plus 25 cents for each mile driven to rent the same car. How many miles must be driven in a week to make the rental cost for Basic Rental a better deal than Continental's?

Let $x =$ number of miles driven in a week.

$$\overbrace{260}^{\text{Cost for Basic Rental}} < \overbrace{80+0.25x}^{\text{Cost for Continental}}$$

$$260 < 80 + 0.25x$$

$$180 < 0.25x$$

$$\frac{180}{0.25} < \frac{0.25x}{0.25}$$

$$720 < x$$

$$x > 720$$

Driving more than 720 miles per week makes Basic Rental a better deal.

3c. An elevator at a construction site has a maximum capacity of 3000 pounds. If the elevator operator weighs 245 pounds and each cement bag weighs 95 pounds, how many bags of cement can be safely lifted on the elevator in one trip?

Objective #4: Recognize inequalities with no solution or all real numbers as solutions.

✔ *Solved Problem #4*

4a. Solve the inequality: $3(x+1) > 3x+2$

$$3(x+1) > 3x+2$$

$$3x+3 > 3x+2$$

$$3 > 2$$

This expression is always true.

The solution set is $\mathbb{R}$ or $(-\infty, \infty)$.

4b. Solve the inequality: $x+1 \leq x-1$

$$x+1 \leq x-1$$

$$1 \leq -1$$

This expression is always false.

The solution set is $\varnothing$.

Pencil Problem #4

4a. Solve the inequality: $4(3x-2) - 3x < 3(1+3x) - 7$

4b. Solve the inequality: $5(x-2) - 3(x+4) \geq 2x - 20$

Objective #5: Solve compound inequalities.

✔ *Solved Problem #5*

5. Solve the compound inequality: $1 \le 2x+3 < 11$

$$1 \le 2x+3 < 11$$
$$1-3 \le 2x+3-3 < 11-3$$
$$-2 \le 2x < 8$$
$$\frac{-2}{2} \le \frac{2x}{2} < \frac{8}{2}$$
$$-1 \le x < 4$$

The solution set is $[-1,4)$.

Pencil Problem #5

5. Solve the compound inequality: $-11 < 2x-1 \le -5$

Objective #6: Solve absolute value inequalities.

✔ *Solved Problem #2*

6a. Solve the inequality:

$$|x-2| < 5$$

Rewrite without absolute value bars.
$|u| < c$ means $-c < u < c$.

$$-5 < x-2 < 5$$
$$-5+2 < x-2+2 < 5+2$$
$$-3 < x < 7$$

The solution set is $(-3,7)$.

6b. Solve the inequality:

$$-3|5x-2|+20 \ge -19$$

First, isolate the absolute value expression on one side of the inequality.

$$-3|5x-2|+20 \ge -19$$
$$-3|5x-2| \ge -39$$
$$\frac{-3|5x-2|}{-3} \le \frac{-39}{-3}$$
$$|5x-2| \le 13$$

(Continued on next page)

Pencil Problem #2

6a. Solve the inequality:

$$|2x-6| < 8$$

6b. Solve the inequality:

$$|2(x-1)+4| \le 8$$

Rewrite $|5x-2| \le 13$ without absolute value bars.

$|u| \le c$ means $-c \le u \le c$.

$$-13 \le 5x - 2 \le 13$$
$$-13+2 \le 5x-2+2 \le 13+2$$
$$-11 \le 5x \le 15$$
$$\frac{-11}{5} \le \frac{5x}{5} \le \frac{15}{5}$$
$$-\frac{11}{5} \le x \le 3$$

The solution set is $\left[\frac{-11}{5}, 3\right]$.

6c. Solve the inequality: $18 < |6-3x|$

Rewrite with the absolute value expression on the left.

$|6-3x| > 18$

This means the same as $6-3x < -18$ or $6-3x > 18$.

$$6-3x < -18 \quad \text{or} \quad 6-3x > 18$$
$$-3x < -24 \qquad -3x > 12$$
$$\frac{-3x}{-3} > \frac{-24}{-3} \qquad \frac{-3x}{-3} < \frac{12}{-3}$$
$$x > 8 \qquad x < -4$$

The solution set is $\{x \mid x < -4 \text{ or } x > 8\}$ or $(-\infty, -4) \cup (8, \infty)$.

6c. Solve the inequality: $1 < |2-3x|$

Answers for Pencil Problems *(Textbook Exercise references in parentheses)*:

1a. $\{x \mid 1 < x \le 6\}$; *(1.7 #1)*

1b. $\{x \mid x \ge -3\}$; *(1.7 #9)*

2a. $[-1, 0)$ (1.7 #15) **2b.** $(-3, 2]$ *(1.7 #17)*

3a. $(-\infty, -4]$; *(1.7 #31)*

3b. $[-10, \infty)$; *(1.7 #41)* **3c.** at most 29 bags *(1.7 #129)*

4a. $(-\infty, \infty)$ *(1.7 #47)* **4b.** $\varnothing$ *(1.7 #49)*

5. $(-5, -2]$ *(1.7 #55)*

6a. $(-1, 7)$ *(1.7 #63)* **6b.** $[-5, 3]$ *(1.7 #65)* **6c.** $\left(-\infty, \frac{1}{3}\right) \cup (1, \infty)$ *(1.7 #89)*

Section 2.1
Basics of Functions and Their Graphs

Say *WHAT???*

You may have noticed that mathematical notation occasionally can have more than one meaning depending on the context.

For example, $(-3, 6)$ could refer to the ordered pair where $x = -3$ and $y = 6$, or it could refer to the open interval $-3 < x < 6$.

Similarly, in this section of the textbook, we will use the notation, $f(x)$. It may surprise you to find out that it does *not* mean to multiply "f times x."

It will be important for you to gain an understanding of what this notation *does* mean as you work through this essential concept of "functions."

Objective #1: Find the domain and range of a relation.

✔ *Solved Problem #1*

1. Find the domain and range of the relation:
   $\{(0, 9.1), (10, 6.7), (20, 10.7), (30, 13.2), (40, 21.2)\}$

The domain is the set of all first components.

Domain:
$\{0, 10, 20, 30, 40\}$.

The range is the set of all second components.

Range:
$\{9.1, 6.7, 10.7, 13.2, 21.2\}$.

Pencil Problem #1

1. Find the domain and range of the relation:
   $\{(3, 4), (3, 5), (4, 4), (4, 5)\}$

Objective #2: Determine whether a relation is a function.

✔ *Solved Problem #2*

2a. Determine whether the relation is a function:
$\{(1,2), (3,4), (5,6), (5,8)\}$

5 corresponds to both 6 and 8. If any element in the domain corresponds to more than one element in the range, the relation is not a function.

Thus, the relation is not a function.

2a. Determine whether the relation is a function:
$\{(3, 4), (3, 5), (4, 4), (4, 5)\}$

2b. Determine whether the relation is a function:

$$\{(1,2),(3,4),(6,5),(8,5)\}$$

Every element in the domain corresponds to exactly one element in the range. No two ordered pairs in the given relation have the same first component and different second components.

Thus, the relation is a function.

2b. Determine whether the relation is a function:

$$\{(-3,-3),(-2,-2),(-1,-1),(0,0)\}$$

Objective #3: Determine whether an equation represents a function.

Solved Problem #3

3. Solve each equation for y and then determine whether the equation defines y as a function of x.

3a. $2x+y=6$

Subtract $2x$ from both sides to solve for y.

$$2x+y=6$$
$$2x-2x+y=6-2x$$
$$y=6-2x$$

For each value of x, there is only one value of y, so the equation defines y as a function of x.

3b. $x^2+y^2=1$

Subtract x^2 from both sides and then use the square root property to solve for y.

$$x^2+y^2=1$$
$$x^2-x^2+y^2=1-x^2$$
$$y^2=1-x^2$$
$$y=\pm\sqrt{1-x^2}$$

For values of x between -1 and 1, there are two values of y. For example, if $x=0$, then $y=\pm 1$. Thus, the equation does not define y as a function of x.

Pencil Problem #3

3. Solve each equation for y and then determine whether the equation defines y as a function of x.

3a. $x^2+y=16$

3b. $x=y^2$

Objective #4: Evaluate a function.

✔ *Solved Problem #4*

4. If $f(x) = x^2 - 2x + 7$, evaluate each of the following.

4a. $f(-5)$

Substitute −5 for x. Place parentheses around −5 when making the substitution.

$$f(-5) = (-5)^2 - 2(-5) + 7$$
$$= 25 + 10 + 7 = 42$$

4b. $f(x+4)$

Substitute $x + 4$ for x and then simplify. Place parentheses around $x + 4$ when making the substitution.

Use $(A+B)^2 = A^2 + 2AB + B^2$ to expand $(x+4)^2$ and the distributive property to multiply $-2(x+4)$. Then combine like terms.

$$f(x+4) = (x+4)^2 - 2(x+4) + 7$$
$$= x^2 + 8x + 16 - 2x - 8 + 7$$
$$= x^2 + 6x + 15$$

4c. $f(-x)$

Substitute $-x$ for x. Place parentheses around $-x$ when making the substitution.

$$f(-x) = (-x)^2 - 2(-x) + 7$$
$$= x^2 + 2x + 7$$

Pencil Problem #4

4. If $g(x) = x^2 + 2x + 3$, evaluate each of the following.

4a. $g(-1)$

4b. $g(x+5)$

4c. $g(-x)$

Objective #5: Graph functions by plotting points.

✔ *Solved Problem #5*

5. Graph the functions $f(x) = 2x$ and $g(x) = 2x - 3$ in the same rectangular coordinate system. Select integers for x, starting with -2 and ending with 2. How is the graph of g related to the graph of f?

Make a table for $f(x) = 2x$:

x	$f(x) = 2x$	(x, y)
-2	$f(-2) = 2(-2) = -4$	$(-2, -4)$
-1	$f(-1) = 2(-1) = -2$	$(-1, -2)$
0	$f(0) = 2(0) = 0$	$(0, 0)$
1	$f(1) = 2(1) = 2$	$(1, 2)$
2	$f(2) = 2(2) = 4$	$(2, 4)$

Make a table for $g(x) = 2x - 3$:

x	$g(x) = 2x - 3$	(x, y)
-2	$g(-2) = 2(-2) - 3 = -7$	$(-2, -7)$
-1	$g(-1) = 2(-1) - 3 = -5$	$(-1, -5)$
0	$g(0) = 2(0) - 3 = -3$	$(0, -3)$
1	$g(1) = 2(1) - 3 = -1$	$(1, -1)$
2	$g(2) = 2(2) - 3 = 1$	$(2, 1)$

Plot the points and draw the lines that pass through them.

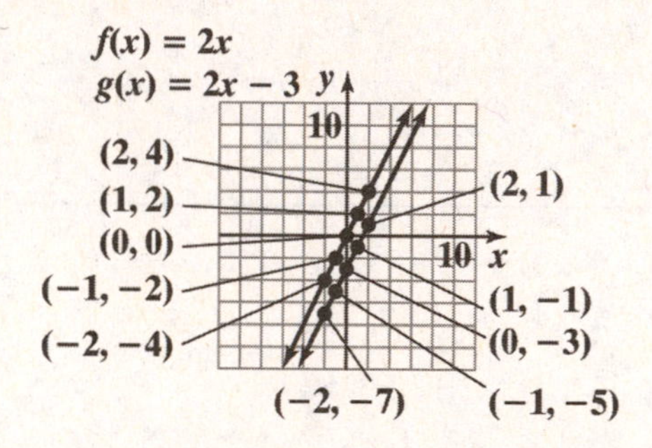

The graph of g is the graph of f shifted down by 3 units.

✎ *Pencil Problem #5*

5. Graph the functions $f(x) = |x|$ and $g(x) = |x| - 2$ in the same rectangular coordinate system. Select integers for x, starting with -2 and ending with 2. How is the graph of g related to the graph of f?

Objective #6: Use the vertical line test to identify functions.

✔ Solved Problem #6

6. Use the vertical line test to determine if the graph represents y as a function of x.

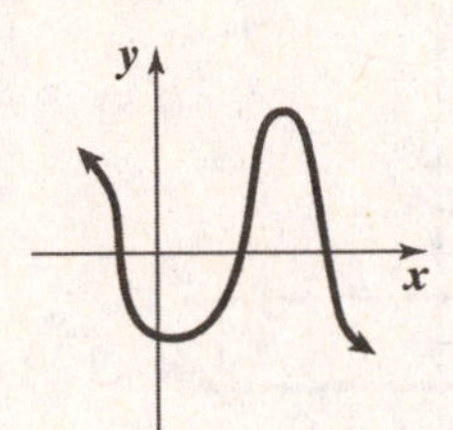

The graph passes the vertical line test and thus y is a function of x.

Pencil Problem #6

6. Use the vertical line test to determine if the graph represents y as a function of x.

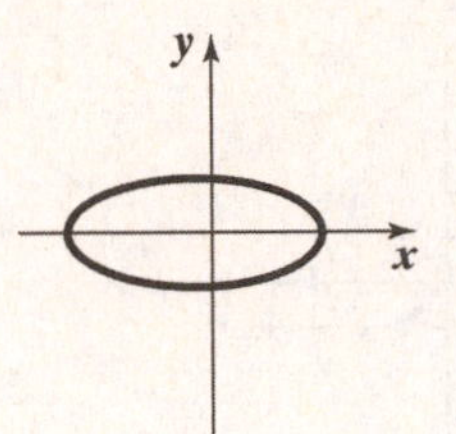

Objective #7: Obtain information about a function from its graph.

✔ Solved Problem #7

7a. The following is the graph of g.

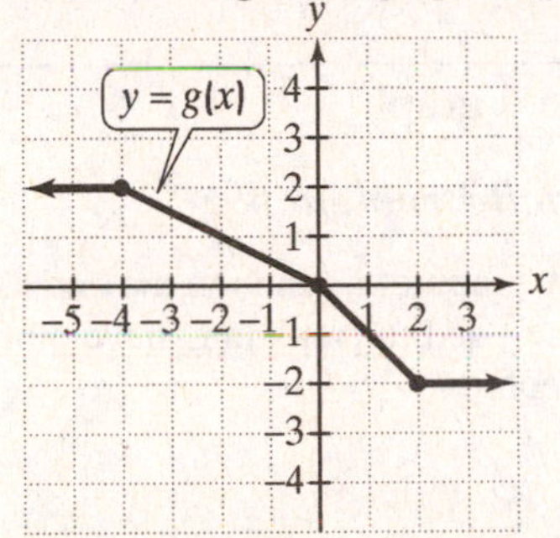

Use the graph to find $g(-20)$.

The graph indicates that to the left of $x = -4$, the graph is at a constant height of 2.

Thus, $g(-20) = 2$.

7b. Use the graph from *Problem 7a* above to find the value of x for which $g(x) = -1$.

$g(1) = -1$

The height of the graph is -1 when $x = 1$.

Pencil Problem #7

7a. The following is the graph of g.

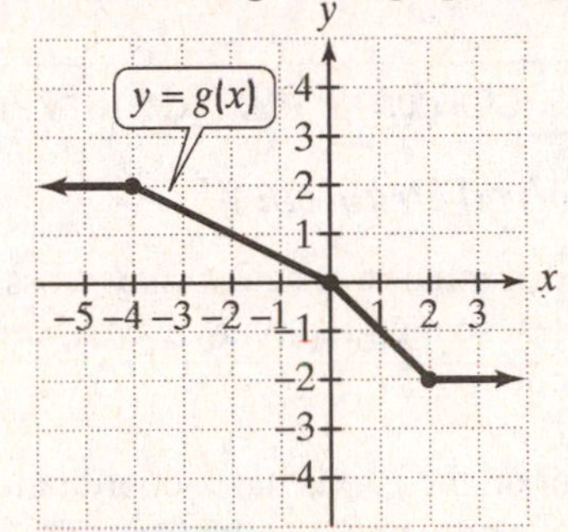

Use the graph to find $g(-4)$.

7b. Use the graph from *Problem 7a* above to find the value of x for which $g(x) = 1$.

Objective #8: Identify the domain and range of a function from its graph.

✔ *Solved Problem #8*

8. Use the graph of the function to identify its domain and its range.

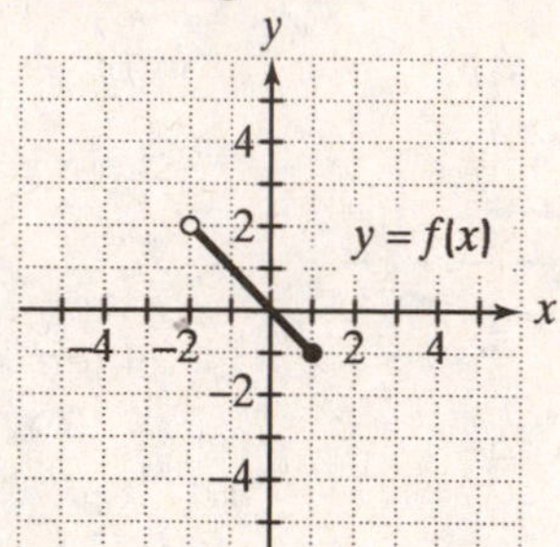

Inputs on the x-axis extend from –2, excluding −2, to 1, including 1.
The domain is $(-2,1]$.

Outputs on the y-axis extend from −1, including −1, to 2, excluding 2.
The range is $[-1,2)$.

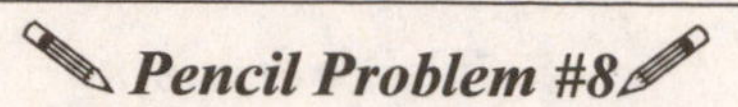

Pencil Problem #8

8. Use the graph of the function to identify its domain and its range.

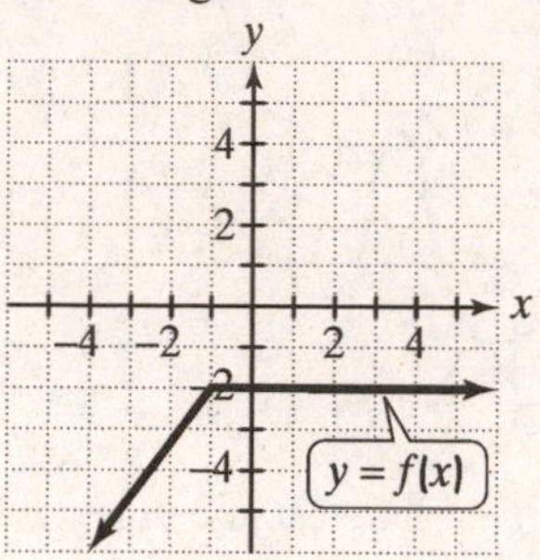

Objective #9: Identify intercepts from a function's graph.

✔ *Solved Problem #9*

9. True or false: The graph of a function may cross the y-axis several times, so the graph may have more than one y-intercept.

False. Since each point on the y-axis has x-coordinate 0 and a function may have only one y-value for each x-value, the graph of a function has at most one y-coordinate.

Pencil Problem #9

9. True or false: The graph of a function may cross the x-axis several times, so the graph may have more than one x-intercept.

Answers for Pencil Problems *(Textbook Exercise references in parentheses)*:

1. Domain: {3, 4}. Range: {4, 5}. *(2.1 #3)* **2a.** not a function *(2.1 #3)* **2b.** function *(2.1 #7)*

3a. $y = 16 - x^2$; y is a function of x. *(2.1 #13)* **3b.** $y = \pm\sqrt{x}$; y is not a function of x. *(2.1 #17)*

4a. 2 *(2.1 #29a)* **4b.** $x^2 + 12x + 38$ *(2.1 #29b)* **4c.** $x^2 - 2x + 3$ *(2.1 #29c)*

5.

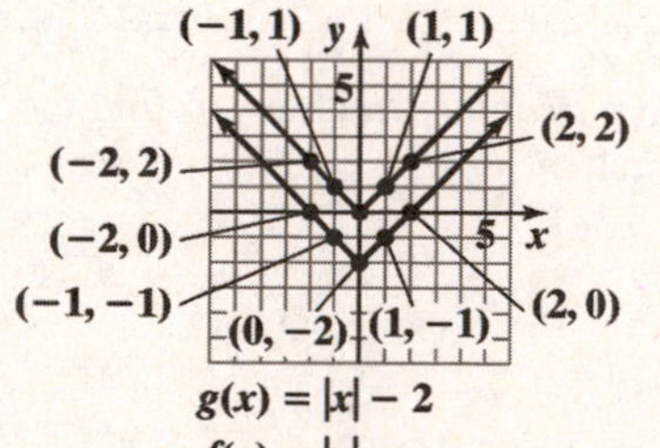

The graph of g is the graph of f shifted down 2 units. *(2.1 #45)*

6. not a function *(2.1 #59)* **7a.** 2 *(2.1 #71)* **7b.** −2 *(2.1 #75)*

8. Domain: $(-\infty,\infty)$. Range: $(-\infty,-2]$ *(2.1 #87)* **9.** True *(2.1 #77)*

Section 2.2
More on Functions and Their Graphs

Can I Really Tell That from a Graph?

Graphs provide a visual representation of how a function changes over time. Many characteristics of a function are much more evident from the function's graph than from the equation that defines the function.

We can use graphs to determine for what years the fuel efficiency of cars was increasing and decreasing and at what age men and women attain their maximum percent body fat. A graph can even help you understand your cellphone bill better.

Objective #1: Identify intervals on which a function increases, decreases, or is constant.

✔ *Solved Problem #1*

1. State the intervals on which the given function is increasing, decreasing, or constant.

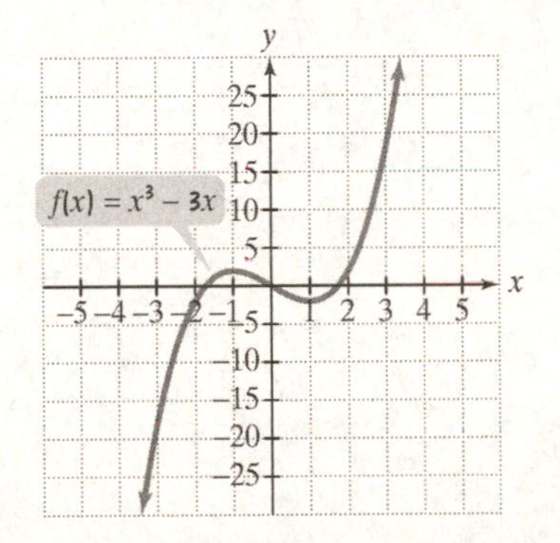

The intervals are stated in terms of x-values.
When we start at the left and follow along the graph, at first the graph is going up. This continues until $x = -1$. The function is increasing on the interval $(-\infty, -1)$.

At $x = -1$, the graph turns and moves downward until we get to $x = 1$. The function is decreasing on the interval $(-1, 1)$.

At $x = 1$, the graph turns again and continues in an upward direction. The function is increasing on the interval $(1, \infty)$.

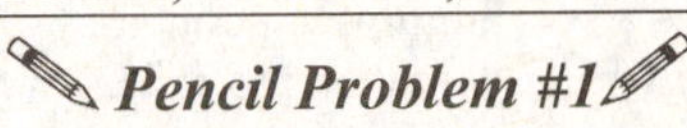

Pencil Problem #1

1. State the intervals on which the given function is increasing, decreasing, or constant.

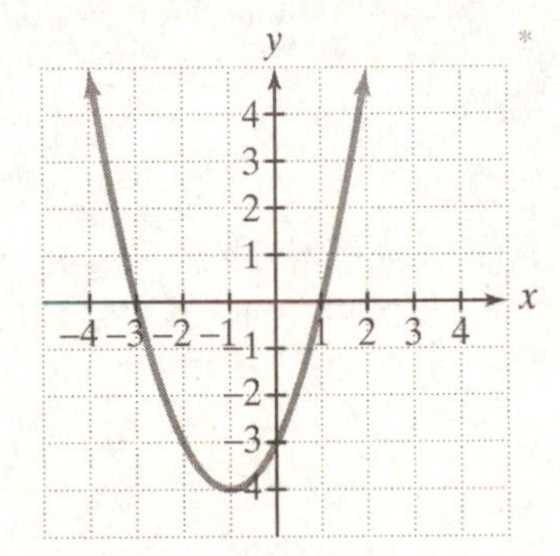

Objective #2: Use graphs to locate relative maxima or minima.

✔ *Solved Problem #2*

2. Look at the graph in Solved Problem #1. Locate values at which the function f has any relative maxima or minima. What are these relative maxima or minima?

The graph has a turning point at $x = -1$. The value of $f(x)$ or y at $x = -1$ is greater than the values of $f(x)$ for values of x near -1 (for values of x between -2 and 0, for example). Thus, f has a relative maximum at $x = -1$. The relative maximum is the value of $f(x)$ or y corresponding to $x = -1$. Using the equation in the graph, we find that $f(-1) = (-1)^3 - 3(-1) = 2$. We say that f has a relative maximum of 2 at $x = -1$.

The graph has a second turning point at $x = 1$. The value of $f(x)$ or y at $x = 1$ is less than the values of $f(x)$ for values of x near 1 (for values of x between 0 and 2, for example). Thus, f has a relative minimum at $x = 1$. The relative minimum is the value of $f(x)$ or y corresponding to $x = 1$. Using the equation in the graph, we find that $f(1) = (1)^3 - 3(1) = -2$. We say that f has a relative minimum of -2 at $x = 1$.

Note that the relative maximum occurs where the functions changes from increasing to decreasing and the relative minimum occurs where the graph changes from decreasing to increasing.

Pencil Problem #2

2. The graph of a function f is given below. Locate values at which the function f has any relative maxima or minima. What are these relative maxima or minima? Read y-values from the graph, as needed, since the equation is not given.

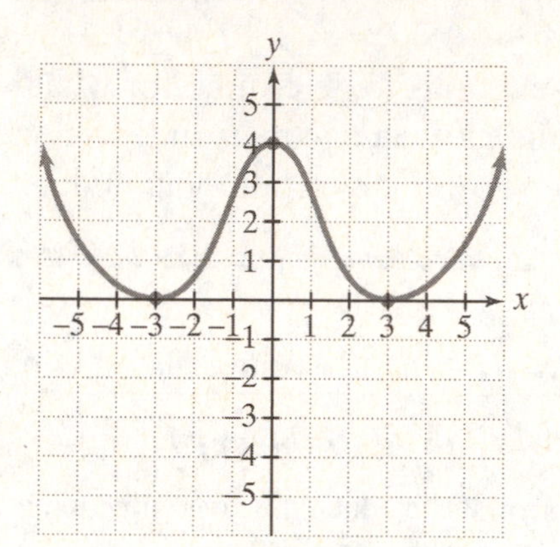

Objective #3: Identify even or odd functions and recognize their symmetries.

✔ *Solved Problem #3*

3. Determine whether each of the following functions is even, odd, or neither.

3a. $f(x) = x^2 + 6$

Replace x with $-x$.

$f(-x) = (-x)^2 + 6 = x^2 + 6 = f(x)$

The function did not change when we replaced x with $-x$. The function is even.

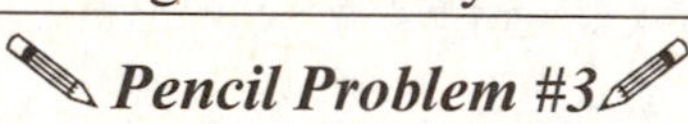

Pencil Problem #3

3. Determine whether each of the following functions is even, odd, or neither.

3a. $f(x) = x^3 + x$

3b. $g(x) = 7x^3 - x$

Replace x with $-x$.
$g(-x) = 7(-x)^3 - (-x) = -7x^3 + x = -g(x)$

Each term of the equation defining the function changed sign when we replaced x with $-x$. The function is odd.

3b. $g(x) = x^2 + x$

3c. $h(x) = x^5 + 1$

Replace x with $-x$.
$h(x) = (-x)^5 + 1 = -x^5 + 1$

The resulting function is not equal to the original function, so the function is not even. Only the sign of one term changed, so the function is not odd. The function is neither even nor odd.

3c. $h(x) = x^2 - x^4$

Objective #4: Understand and use piecewise functions.

✔ *Solved Problem #4*

4. Graph the piecewise function defined by

$$f(x) = \begin{cases} 3 & \text{if } x \le -1 \\ x - 2 & \text{if } x > -1 \end{cases}$$

For $x \le -1$, the function value is always 3, so (–4, 3) and (–1, 3) are examples of points on the first piece of the graph.

For $x > -1$, we use $f(x) = x - 2$. We have points such as (0, –2) and (2, 0) on the graph. Note that this piece of the graph will approach the point (–1, –3) but this point is not part of the graph. We use an open dot at (–1, –3).

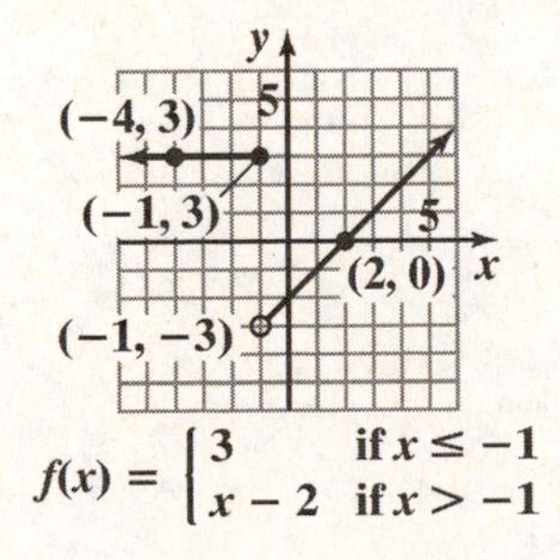

4. Graph the piecewise function defined by

$$f(x) = \begin{cases} 2x & \text{if } x \le 0 \\ 2 & \text{if } x > 0 \end{cases}$$

Objective #5: Find and simplify a function's difference quotient.

✔ Solved Problem #5

5. Find and simplify the difference quotient for $f(x)=-2x^2+x+5$.

$$\frac{f(x+h)-f(x)}{h}$$

$$=\frac{[-2(x+h)^2+(x+h)+5]-(-2x^2+x+5)}{h}$$

$$=\frac{-2(x^2+2xh+h^2)+x+h+5+2x^2-x-5}{h}$$

$$=\frac{-2x^2-4xh-2h^2+x+h+5+2x^2-x-5}{h}$$

$$=\frac{-4xh-2h^2+h}{h}$$

$$=\frac{h(-4x-2h+1)}{h}$$

$$=-4x-2h+1,\ h\neq 0$$

Pencil Problem #5

5. Find and simplify the difference quotient for $f(x)=x^2-4x+3$.

Answers for Pencil Problems *(Textbook Exercise references in parentheses)*:

1. decreasing on $(-\infty, -1)$; increasing on $(-1, \infty)$ *(2.2 #1)*

2. relative minimum of 0 at $x=-3$; relative maximum of 4 at $x=0$; relative minimum of 0 at $x=3$ *(2.2 #13)*

3a. odd *(2.2 #17)* **3b.** neither *(2.2 #19)* **3c.** even *(2.2 #21)*

4. 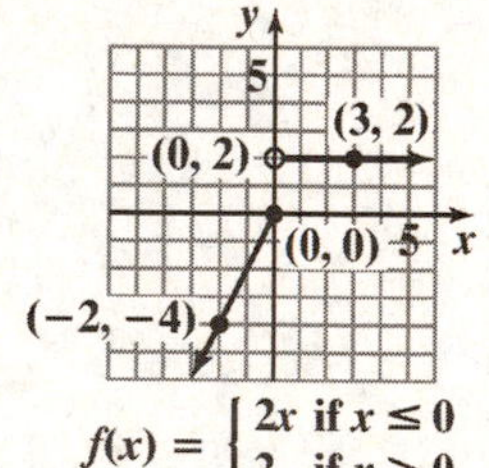

$f(x)=\begin{cases}2x & \text{if } x\le 0\\ 2 & \text{if } x>0\end{cases}$ *(2.2 #45)*

5. $2x+h-4,\ h\neq 0$ *(2.2 #61)*

Section 2.3
Linear Functions and Slope

READ FOR LIFE!

Is there a relationship between literacy and child mortality?

As the percentage of adult females who are literate increases, does the mortality of children under age five decrease? Data from the United Nations indicates that this is, indeed, the case.

In this section of the textbook, you will be given a graph for which each point represents one country. You will use the concept of slope to see how much the mortality rate decreases for each 1% increase in the literacy rate of adult females in a country.

Objective #1: Calculate a line's slope.

✔ *Solved Problem #1*

1. Find the slope of the line passing through $(4,-2)$ and $(-1,5)$.

$$m=\frac{y_2-y_1}{x_2-x_1}$$

$$m=\frac{5-(-2)}{-1-4}$$

$$=\frac{7}{-5}$$

$$=-\frac{7}{5}$$

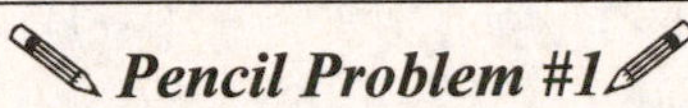

Pencil Problem #1

1. Find the slope of the line passing through $(-2,1)$ and $(2,2)$.

Objective #2: Write the point-slope form of the equation of a line.

✔ *Solved Problem #2*

2a. Write the point-slope form of the equation of the line with slope 6 that passes through the point $(2,-5)$. Then solve the equation for y.

Begin by finding the point-slope equation of a line.

$$y-y_1=m(x-x_1)$$

$$y-(-5)=6(x-2)$$

$$y+5=6(x-2)$$

Now solve this equation for y.

$$y+5=6(x-2)$$

$$y+5=6x-12$$

$$y=6x-17$$

Pencil Problem #2

2a. Write the point-slope form of the equation of the line with slope -3 that passes through the point $(-2,-3)$. Then solve the equation for y.

2b. A line passes through the points $(-2,-1)$ and $(-1,-6)$. Find the equation of the line in point-slope form and then solve the equation for y.

Begin by finding the slope: $m=\dfrac{-6-(-1)}{-1-(-2)}=\dfrac{-5}{1}=-5$

Using the slope and either point, find the point-slope equation of a line.

$$y-y_1=m(x-x_1) \quad \text{or} \quad y-y_1=m(x-x_1)$$
$$y-(-1)=-5(x-(-2)) \quad\quad y-(-6)=-5(x-(-1))$$
$$y+1=-5(x+2) \quad\quad y+6=-5(x+1)$$

To obtain slope-intercept form, solve the above equation for y:

$$y+1=-5(x+2) \quad \text{or} \quad y+6=-5(x+1)$$
$$y+1=-5x-10 \quad\quad y+6=-5x-5$$
$$y=-5x-11 \quad\quad y=-5x-11$$

2b. A line passes through the points $(-3,-1)$ and $(2,4)$. Find the equation of the line in point-slope form and then solve the equation for y.

Objective #3: Write and graph the slope-intercept form of the equation of a line.

✔ *Solved Problem #3*

3. Graph: $f(x)=\dfrac{3}{5}x+1$

The y-intercept is 1, so plot the point (0,1).

The slope is $m=\dfrac{3}{5}$.

Find another point by going up 3 units and to the right 5 units.

Use a straightedge to draw a line through the two points.

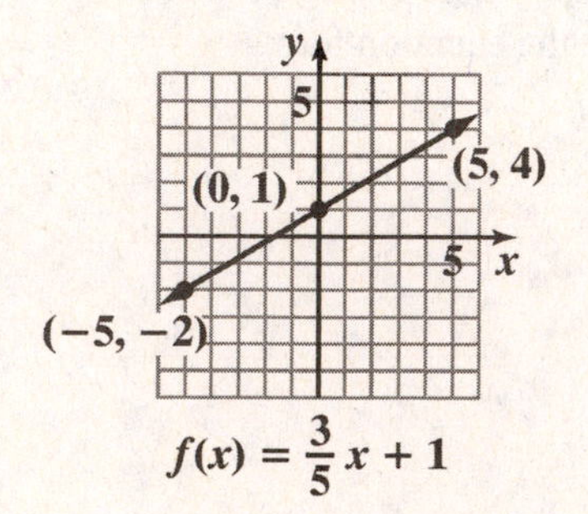

Pencil Problem #3

3. Graph: $f(x)=\dfrac{3}{4}x-2$

Objective #4: Graph horizontal or vertical lines.

✔ *Solved Problem #4*

4a. Graph $y = 3$ in the rectangular coordinate system.

$y = 3$ is a horizontal line.

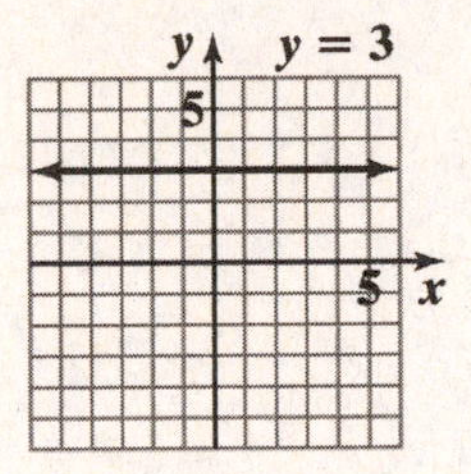

Pencil Problem #4

4a. Graph $y = -2$ in the rectangular coordinate system.

✔ *Solved Problem #4*

4b. Graph $x = -3$ in the rectangular coordinate system.

$x = -3$ is a vertical line.

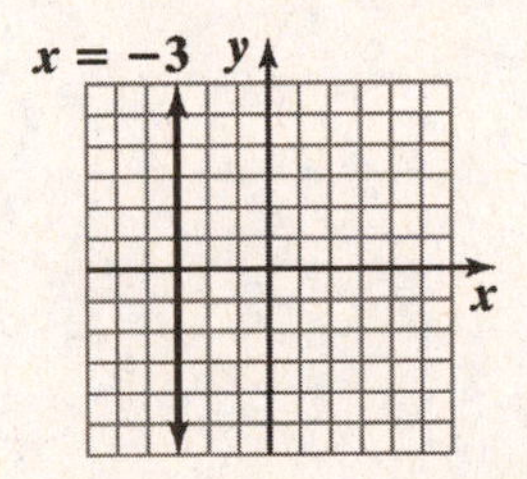

Pencil Problem #4

4b. Graph $x = 5$ in the rectangular coordinate system.

Objective #5: Recognize and use the general form of a line's equation.

✔ *Solved Problem #5*

5. Find the slope and y-intercept of the line whose equation is $3x+6y-12=0$.

Solve for y.

$$3x+6y-12=0$$
$$6y=-3x+12$$
$$\frac{6y}{6}=\frac{-3x+12}{6}$$
$$y=\frac{-3}{6}x+\frac{12}{6}$$
$$y=-\frac{1}{2}x+2$$

The coefficient of x, $-\frac{1}{2}$, is the slope, and the constant term, 2, is the y-intercept.

Pencil Problem #5

5. Find the slope and y-intercept of the line whose equation is $2x+3y-18=0$.

Objective #6: Use intercepts to graph a linear function in standard form.

✔ *Solved Problem #6*

6. Graph: $3x-2y-6=0$

Find the x–intercept by setting $y=0$.

$$3x-2y-6=0$$
$$3x-2(0)-6=0$$
$$3x=6$$
$$x=2$$

Find the y–intercept by setting $x=0$.

$$3x-2y-6=0$$
$$3(0)-2y-6=0$$
$$-2y=6$$
$$y=-3$$

Plot the points and draw the line that passes through them.

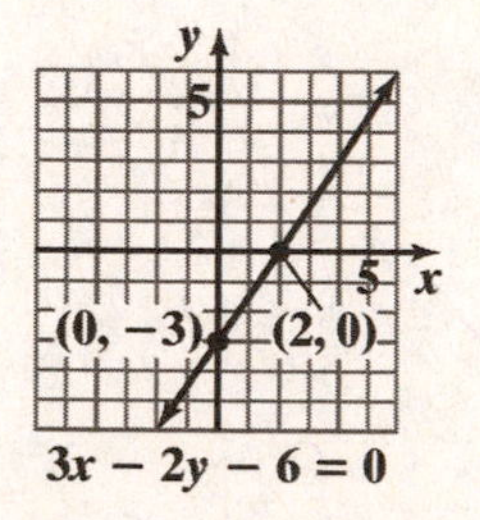

Pencil Problem #6

6. Graph: $6x-2y-12=0$

Objective #7: Model data with linear functions and make predictions.

✔ *Solved Problem #7*

7. The amount of carbon dioxide in the atmosphere, measured in parts per million, has been increasing as a result of the burning of oil and coal. The buildup of gases and particles is believed to trap heat and raise the planet's temperature. When the atmospheric concentration of carbon dioxide is 317 parts per million, the average global temperature is 57.04°F. When the atmospheric concentration of carbon dioxide is 354 parts per million, the average global temperature is 57.64°F.

Write a linear function that models average global temperature, $f(x)$, for an atmospheric concentration of carbon dioxide of x parts per million. Use the function to project the average global temperature when the atmospheric concentration of carbon dioxide is 600 parts per million.

Write the equation of the line through the points (317, 57.04) and (354, 57.64). First find the slope.

$$m = \frac{y_2 - y_1}{x_2 - x_1} = \frac{57.64 - 57.04}{354 - 317} = \frac{0.6}{37} \approx 0.016$$

Use this slope and the point (317, 57.04) in the point-slope form.

$$\begin{aligned} y - y_1 &= m(x - x_1) \\ y - 57.04 &= 0.016(x - 317) \\ y - 57.04 &= 0.016x - 5.072 \\ y &= 0.016x + 51.968 \end{aligned}$$

Using function notation and rounding the constant, we have

$$f(x) = 0.016x + 52.0$$

To predict the temperature when the atmospheric concentration of carbon dioxide is 600 parts per million, find $f(600)$.

$$f(600) = 0.016(600) + 52.0 = 61.6$$

The model predicts an average global temperature of 61.6°F when the atmospheric concentration of carbon dioxide is 600 parts per million.

✎ *Pencil Problem #7* ✎

7. When the literacy rate for adult females in a country is 0%, the infant mortality rate is 254 (per thousand). When the literacy rate for adult females is 60%, the infant mortality rate is 110. Write a linear function that models child mortality, $f(x)$, per thousand, for children under five in a country where x% of adult women are literate. Use the function to predict the child mortality rate in a country where 80% of adult females are literate.

Answers for Pencil Problems *(Textbook Exercise references in parentheses)*:

1. $\frac{1}{4}$ *(2.3 #3)*

2a. $y+3=-3(x+2); y=-3x-9$ *(2.3 #15)* **2b.** $y+1=1(x+3)$ or $y-4=1(x-2); y=x+2$ *(2.3 #29)*

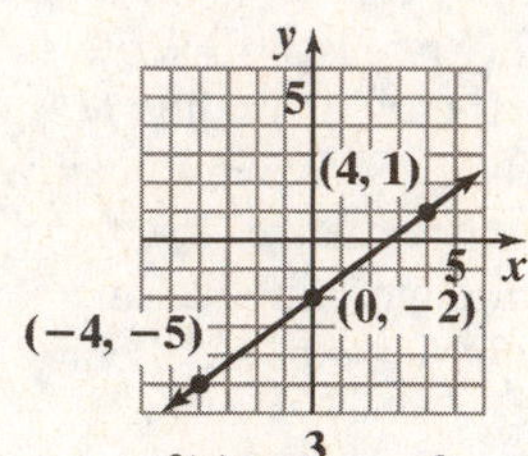

3. *(2.3 #43)*

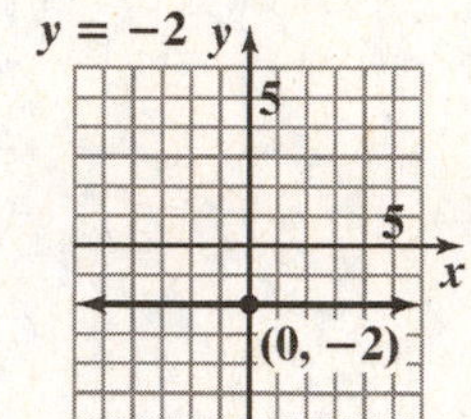

4a. *(2.3 #49)*

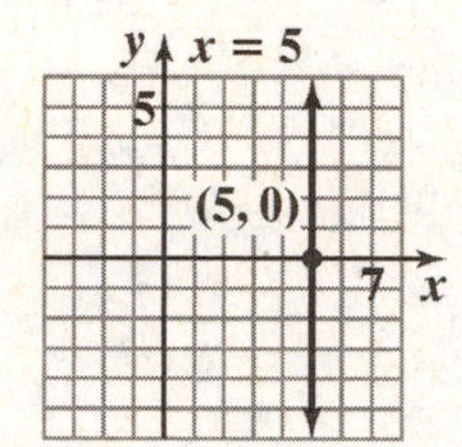

4b. *(2.3 #52)*

5. slope: $-\frac{2}{3}$; y-intercept: 6 *(2.3 #61)*

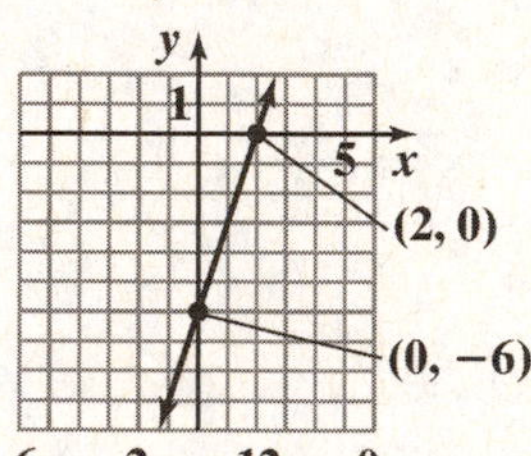

6. *(2.3 #67)*

7. $f(x)=-2.4x+254$; 62 per thousand *(2.3 #71)*

Section 2.4
More on Slope

Will They Ever Catch Up?

Many quantities, such as the number of men and the number of women living alone, are increasing over time. We can use slope to indicate how fast such quantities are growing on average.

If the slopes are the same, the quantities are growing at the same rate. However, if the slopes are different, then one quantity is growing faster than the other.
There were 9.0 million men and 14.0 million women living alone in 1990. Since then the number of men living alone has increased faster than the number of women living alone. If this trend continues, eventually, the number of men living alone will catch up to the number of women living alone.

Objective #1: Find slopes and equations of parallel and perpendicular lines.

✔ *Solved Problem #1*

1a. Write an equation of the line passing through $(-2,5)$ and parallel to the line whose equation is $y = 3x + 1$. Express the equation in point-slope form and slope-intercept form.

Since the line is parallel to $y = 3x + 1$, we know it will have slope $m = 3$.

We are given that it passes through $(-2,5)$. We use the slope and point to write the equation in point-slope form.

$$y - y_1 = m(x - x_1)$$
$$y - 5 = 3(x - (-2))$$
$$y - 5 = 3(x + 2)$$

Point-Slope form: $y - 5 = 3(x + 2)$

Solve for y to obtain slope-intercept form.

$$y - 5 = 3(x + 2)$$
$$y - 5 = 3x + 6$$
$$y = 3x + 11$$
$$f(x) = 3x + 11$$

Slope-Intercept form: $y = 3x + 11$

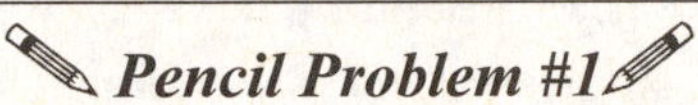

Pencil Problem #1

1a. Write an equation of the line passing through $(-8,-10)$ and parallel to the line whose equation is $y = -4x + 3$. Express the equation in point-slope form and slope-intercept form.

1b. Write an equation of the line passing through $(-2,-6)$ and perpendicular to the line whose equation is $x+3y-12=0$. Express the equation in point-slope form and general form.

First, find the slope of the line $x+3y-12=0$.
Solve the given equation for y to obtain slope-intercept form.

$$x+3y-12=0$$
$$3y=-x+12$$
$$y=-\frac{1}{3}x+4$$

Since the slope of the given line is $-\frac{1}{3}$, the slope of any line perpendicular to the given line is 3.

We use the slope of 3 and the point $(-2,-6)$ to write the equation in point-slope form. Then gather the variable and constant terms on one side with zero on the other side.

$$y-y_1=m(x-x_1)$$
$$y-(-6)=3(x-(-2))$$
$$y+6=3(x+2)$$
$$y+6=3x+6$$
$$0=3x-y \text{ or } 3x-y=0$$

1b. Write an equation of the line passing through $(4,-7)$ and perpendicular to the line whose equation is $x-2y-3=0$. Express the equation in point-slope form and general form.

Objective #2: Interpret slope as rate of change.

✔ *Solved Problem #2*

2. In 1990, there 9.0 million men living alone and in 2008, there were 14.7 million men living alone. Use the ordered pairs (1990, 9.0) and (2008, 14.7) to find the slope of the line through the points. Express the slope correct to two decimal places and describe what it represents.

$$m=\frac{\text{Change in } y}{\text{Change in } x}=\frac{14.7-9.0}{2008-1990}$$
$$=\frac{5.7}{18}\approx 0.32$$

The number of men living alone increased at an average rate of approximately 0.32 million men per year.

Pencil Problem #2

2. In 1994, 617 active-duty servicemembers were discharged under the "don't ask, don't tell" policy. In 1998, 1163 were discharged under the policy. Use the ordered pairs (1994, 617) and (1998, 1163) to find the slope of the line through the points. Express the slope correct to the nearest whole number and describe what it represents.

Objective #3: Find a function's average rate of change..

✔ *Solved Problem #3*

3. Find the average rate of change of the function from x_1 to x_2.

3a. $f(x)=x^3$ from $x_1 = 0$ to $x_2 = 1$

$$\frac{f(x_2)-f(x_1)}{x_2-x_1}=\frac{f(1)-f(0)}{1-0}$$
$$=\frac{1^3-0^3}{1}$$
$$=1$$

The average rate of change is 1.

3b. $f(x)=x^3$ from $x_1 = 1$ to $x_2 = 2$

$$\frac{f(x_2)-f(x_1)}{x_2-x_1}=\frac{f(2)-f(1)}{2-1}$$
$$=\frac{2^3-1^3}{1}$$
$$=\frac{8-1}{1}$$
$$=7$$

The average rate of change is 7.

3c. $f(x)=x^3$ from $x_1 = -2$ to $x_2 = 0$

$$\frac{f(x_2)-f(x_1)}{x_2-x_1}=\frac{f(0)-f(-2)}{0-(-2)}$$
$$=\frac{0^3-(-2)^3}{0+2}$$
$$=\frac{0-(-8)}{2}$$
$$=\frac{8}{2}$$
$$=4$$

The average rate of change is 4.

Pencil Problem #3

3. Find the average rate of change of the function from x_1 to x_2.

3a. $f(x)=3x$ from $x_1 = 0$ to $x_2 = 5$

3b. $f(x)=x^2+2x$ from $x_1 = 3$ to $x_2 = 5$

3c. $f(x)=\sqrt{x}$ from $x_1 = 4$ to $x_2 = 9$

Answers for Pencil Problems *(Textbook Exercise references in parentheses)*:

1a. Point-Slope form: $y+10=-4(x+8)$, Slope-Intercept form: $y=-4x-42$ *(2.4 #5)*

1b. Point-Slope form: $y+7=-2(x-4)$, General form: $2x+y-1=0$ *(2.4 #11)*

2. 137; There was an average increase of approximately 137 discharges per year. *(2.4 #27)*

3a. 3 *(2.4 #13)* **3b.** 10 *(2.4 #15)* **3c.** $\frac{1}{5}$ *(2.4 #17)*

Section 2.5
Transformations of Functions

Movies and Mathematics

Have you ever seen special effects in a movie where a person or object is continuously transformed into something different? This is called morphing.
In mathematics, we can use transformations of a known graph to graph a function with a similar equation. This is achieved through horizontal and vertical shifts, reflections, and stretching and shrinking of the known graph.

Objective #1: Recognize graphs of common functions.

✔ ***Solved Problem #1***

1. True or false: The graphs of the standard quadratic function $f(x) = x^2$ and the absolute value function $g(x) = |x|$ have the same type of symmetry.

True. Both functions are even and their graphs are symmetric with respect to the y-axis.

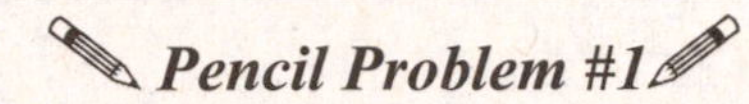

Pencil Problem #1

1. True or false: The graphs of the identity function $f(x) = x$ and the standard cubic function $g(x) = x^3$ have the same type of symmetry.

Objective #2: Use vertical shifts to graph functions.

✔ ***Solved Problem #2***

2. Use the graph of $f(x) = |x|$ to obtain the graph of $g(x) = |x| + 3$.

The graph of g is the graph of f shifted vertically up by 3 units. Add 3 to each y-coordinate.

Since the points (–3, 3), (0, 0), and (3, 3) are on the graph of f, the points (–3, 6), (0, 3), and (3, 6) are on the graph of g.

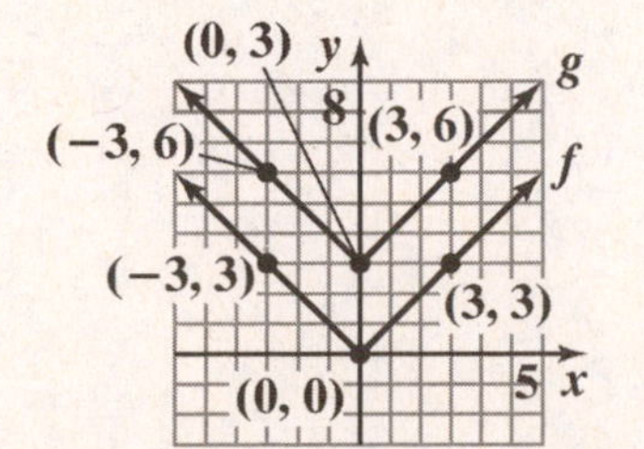

Pencil Problem #2

2. Use the graph of $f(x) = x^2$ to obtain the graph of $g(x) = x^2 - 2$.

Objective #3: Use horizontal shifts to graph functions.

✔ *Solved Problem #3*

3. Use the graph of $f(x) = \sqrt{x}$ to obtain the graph of $g(x) = \sqrt{x-4}$.

$g(x) = \sqrt{x-4} = f(x-4)$

The graph of g is the graph of f shifted horizontally to the right by 4 units. Add 4 to each x-coordinate.

Since the points (0, 0), (1, 1), and (4, 2) are on the graph of f, the points (4, 0), (5, 1), and (8, 2) are on the graph of g.

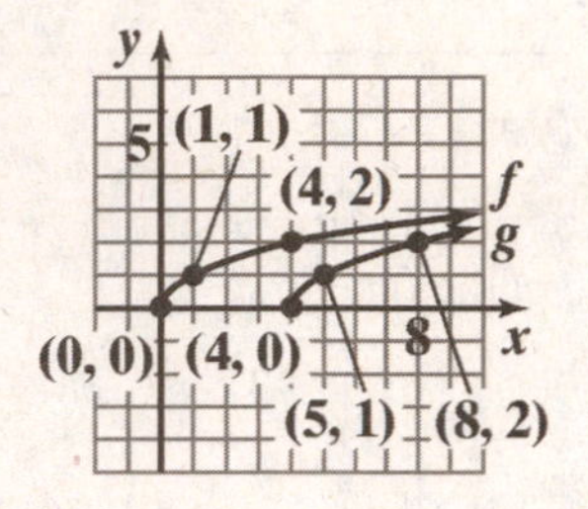

Pencil Problem #3

3. Use the graph of $f(x) = |x|$ to obtain the graph of $g(x) = |x+4|$.

Objective #4: Use reflections to graph functions.

✔ *Solved Problem #4*

4a. Use the graph of $f(x) = |x|$ to obtain the graph of $g(x) = -|x|$.

The graph of g is a reflection of the graph of f about the x-axis because $g(x) = -f(x)$. Replace each y-coordinate with its opposite.

Since the points (–3, 3), (0, 0), and (3, 3) are on the graph of f, the points (–3, –3), (0, 0), and (3, –3) are on the graph of g.

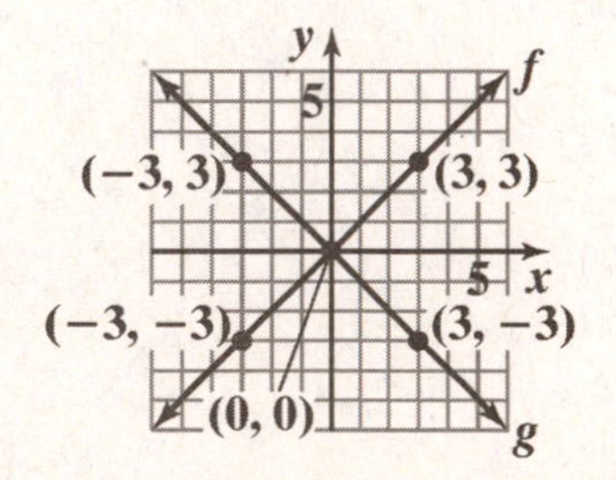

Pencil Problem #4

4a. Use the graph of $f(x) = x^3$ to obtain the graph of $h(x) = -x^3$.

4b. Use the graph of $f(x)=\sqrt[3]{x}$ to obtain the graph of $h(x)=\sqrt[3]{-x}$.

The graph of h is a reflection of the graph of f about the y-axis because $h(x)=f(-x)$. Replace each x-coordinate with its opposite.

Since the points (–1, –1), (0, 0), and (1, 1) are on the graph of f, the points (1, –1), (0, 0), and (–1, 1) are on the graph of h.

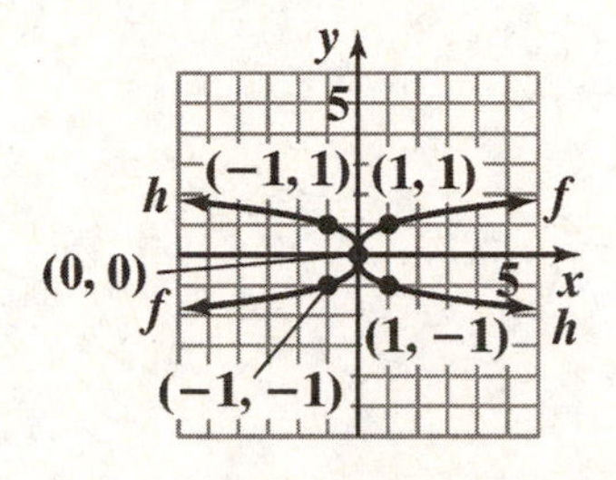

4b. Use the graph of f shown below to obtain the graph of $g(x)=f(-x)$.

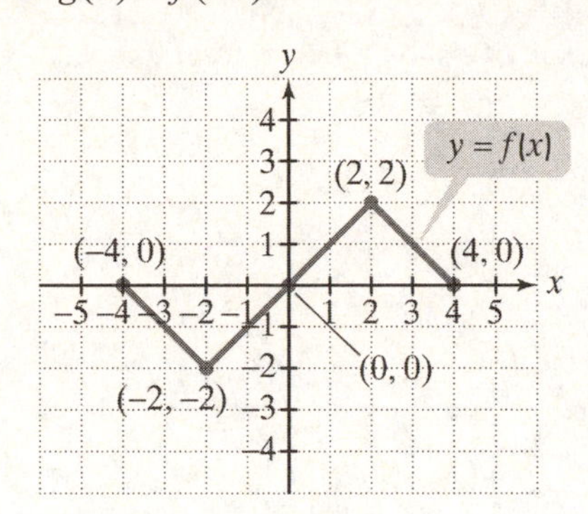

Objective #5: Use vertical stretching and shrinking to graph functions.

Solved Problem #5

5. Use the graph of $f(x)=|x|$ to obtain the graph of $g(x)=2|x|$.

The graph of g is obtained by vertically stretching the graph of f because $g(x)=2f(x)$. Multiply each y-coordinate by 2.

Since the points (–2, 2), (0, 0), and (2, 2) are on the graph of f, the points (–2, 4), (0, 0), and (2, 4) are on the graph of g.

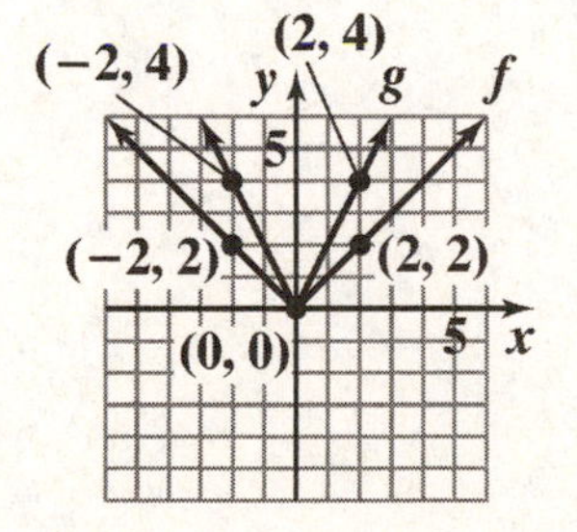

Pencil Problem #5

5. Use the graph of $f(x)=x^3$ to obtain the graph of $h(x)=\frac{1}{2}x^3$.

Objective #6: Use horizontal stretching and shrinking to graph functions.

✔ **Solved Problem #6**

6. Use the graph of f shown below to graph each function.

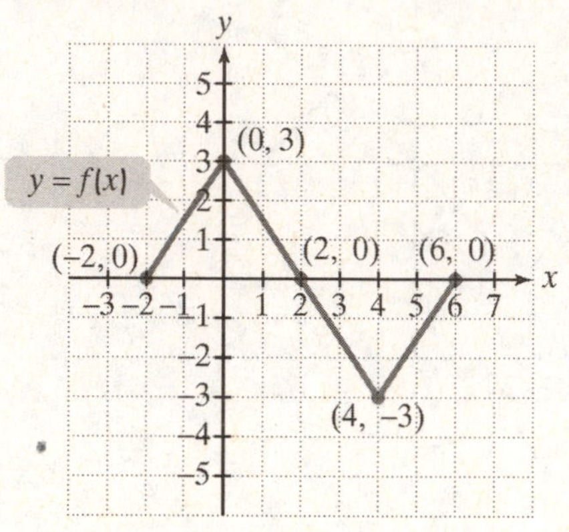

6a. $g(x) = f(2x)$

Divide the x-coordinate of each point on the graph of f by 2. The points $(-1, 0)$, $(0, 3)$, $(1, 0)$, $(2, -3)$, and $(3, 0)$ are on the graph of g.

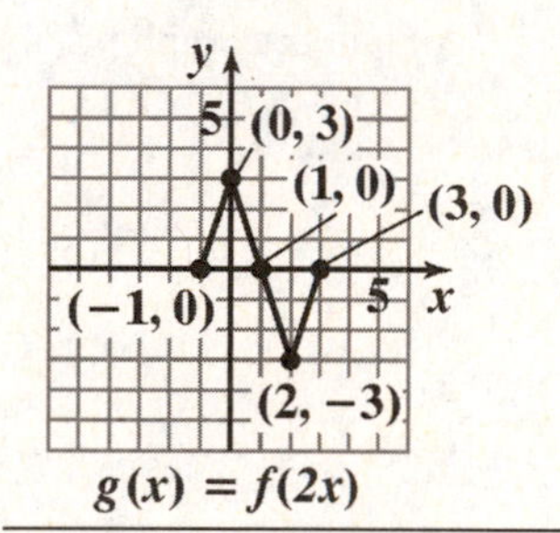

$g(x) = f(2x)$

6b. $h(x) = f(\frac{1}{2}x)$

Multiply the x-coordinate of each point on the graph of f by 2. The points $(-4, 0)$, $(0, 3)$, $(4, 0)$, $(8, -3)$, and $(12, 0)$ are on the graph of h.

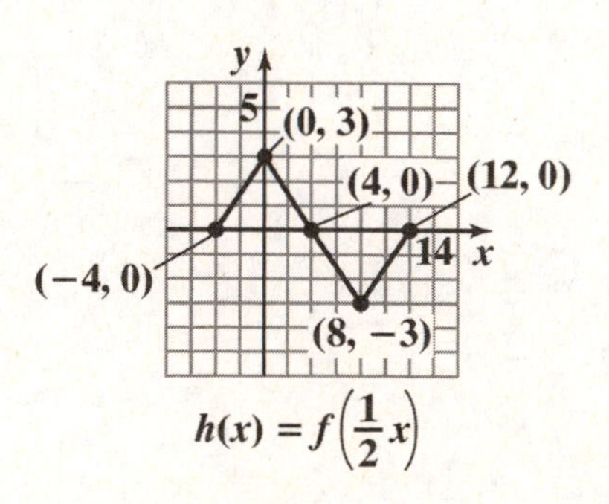

$h(x) = f\left(\frac{1}{2}x\right)$

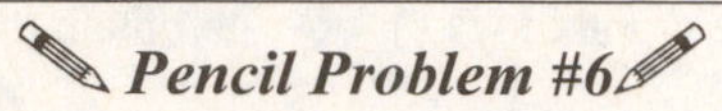

Pencil Problem #6

6. Use the graph of f shown below to graph each function.

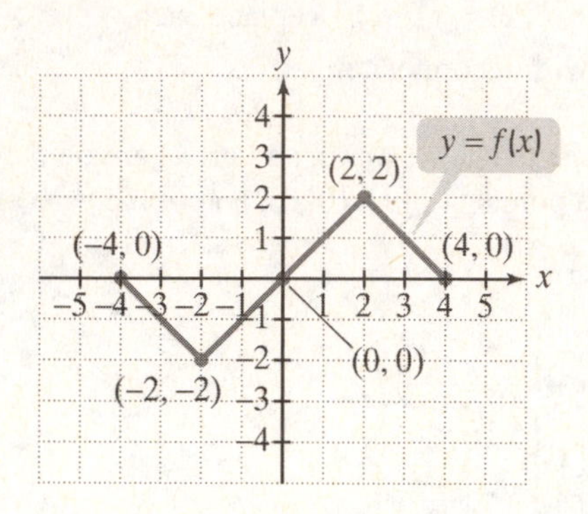

6a. $g(x) = f(2x)$

6b. $g(x) = f(\frac{1}{2}x)$

Objective #7: Graph functions involving a sequence of transformations.

✔ Solved Problem #7

7. Use the graph of $f(x) = x^2$ to graph

$g(x) = 2(x-1)^2 + 3.$

The graph of g is the graph of f horizontally shifted to the right 1 unit, vertically stretched by a factor of 2, and vertically shifted up 3 units. Beginning with a point on the graph of f, add 1 to each x-coordinate, then multiply each y-coordinate by 2, and finally add 3 to each y-coordinate.

$(-1, 1) \to (0, 1) \to (0, 2) \to (0, 5)$
$(0, 0) \to (1, 0) \to (1, 0) \to (1, 3)$
$(1, 1) \to (2, 1) \to (2, 2) \to (2, 5)$

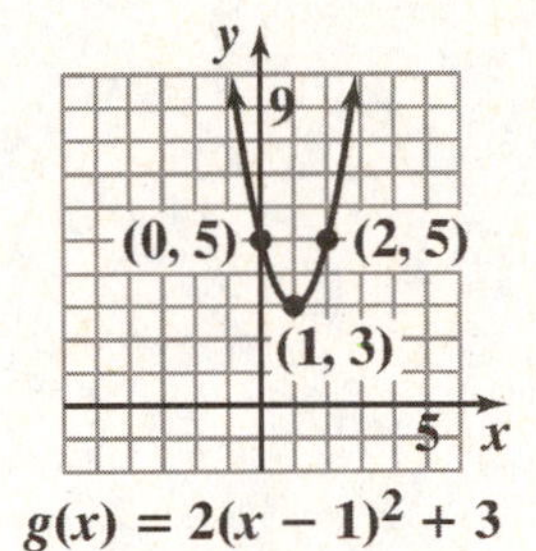

$g(x) = 2(x-1)^2 + 3$

Pencil Problem #7

7. Use the graph of $f(x) = x^3$ to graph

$h(x) = \frac{1}{2}(x-3)^2 - 2.$

Answers for Pencil Problems *(Textbook Exercise references in parentheses)*:

1. True *(2.5 #99)*

2.

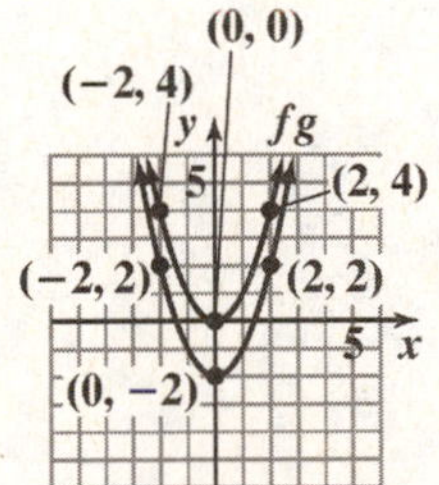

(2.5 #53)

3.

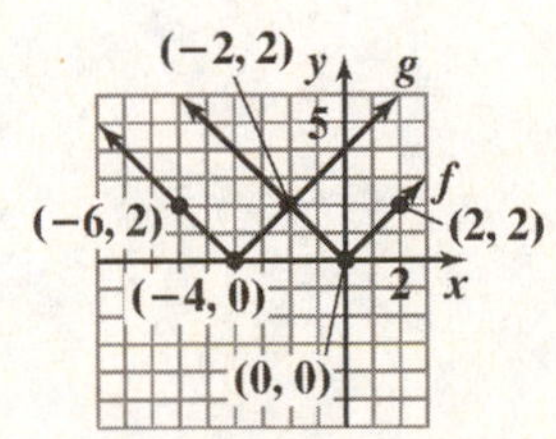

(2.5 #83)

4a.

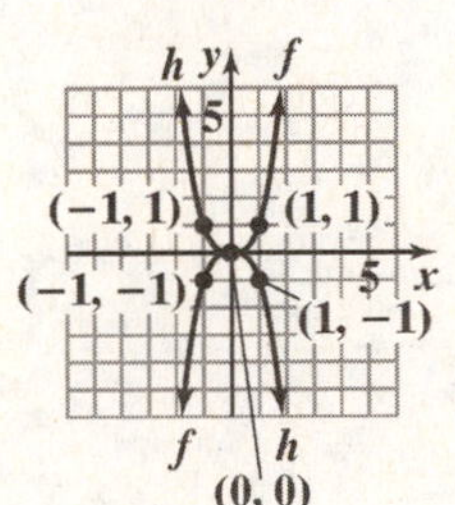

(2.5 #99)

4b.

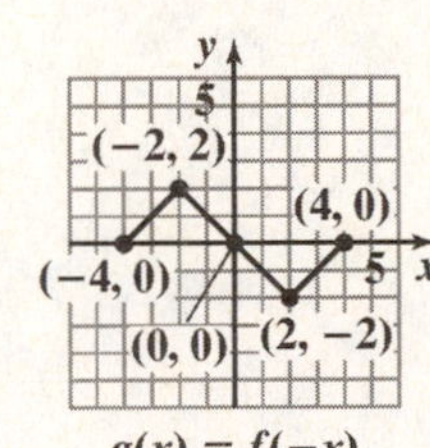

$g(x) = f(-x)$

(2.5 #24)

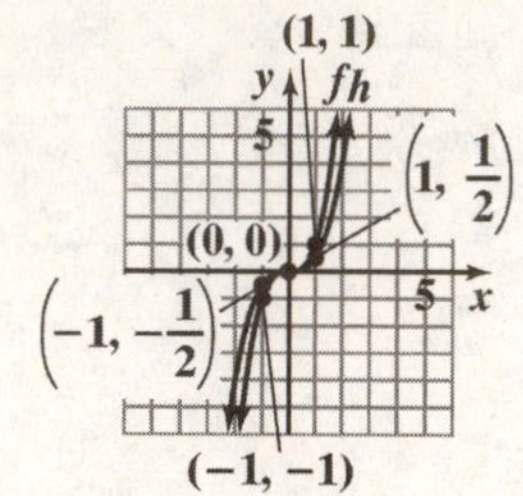

5. *(2.5 #101)*

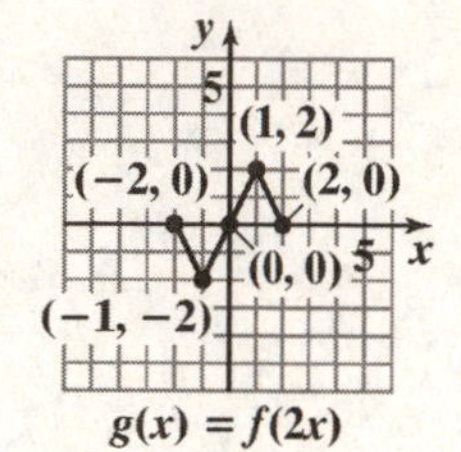

$g(x) = f(2x)$

6a. 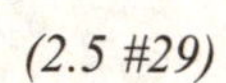 *(2.5 #29)*

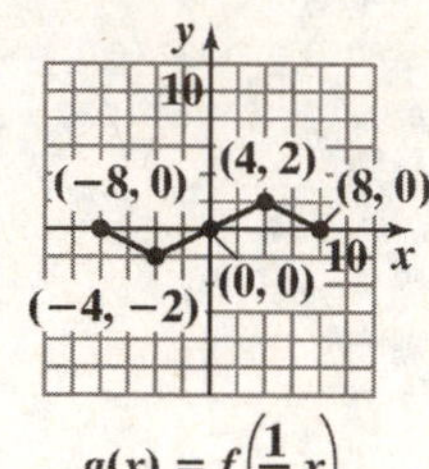

$g(x) = f\left(\frac{1}{2}x\right)$

6b. *(2.5 #30)*

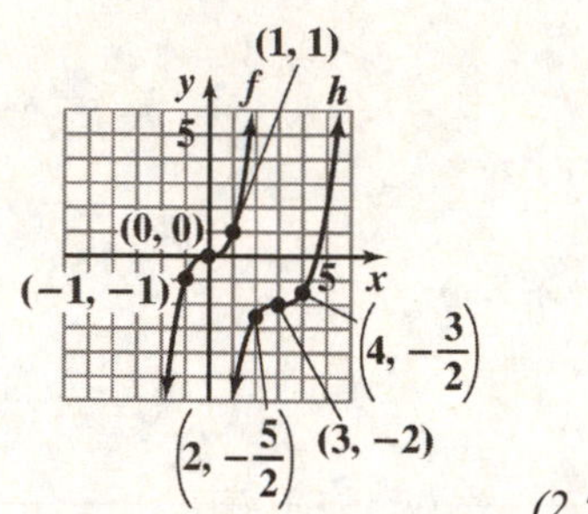

7. *(2.5 #105)*

Section 2.6
Combinations of Functions; Composite Functions

We're Born. We Die.

The figure below quantifies these statements by showing the number of births and deaths in the United States from 2000 through 2009.

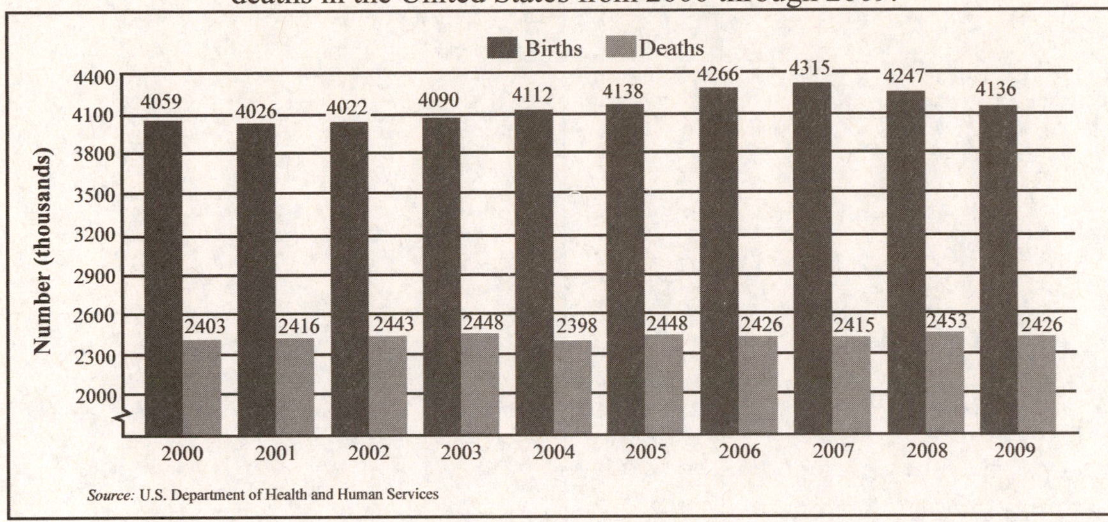

In this section, we look at these data from the perspective of functions. By considering the yearly change in the U.S. population, you will see that functions can be subtracted using procedures that will remind you of combining algebraic expressions.

Objective #1: Find the domain of a function.

✔ *Solved Problem #1*

1a. Find the domain of $f(x)=x^2+3x-17$.

The function contains neither division nor an even root. It is defined for all real numbers. The domain is $(-\infty, \infty)$.

1b. Find the domain of $j(x)=\dfrac{5x}{\sqrt{24-3x}}$.

The function contains both an even root and division. The expression under the radical must be nonnegative and the denominator cannot equal 0. Thus, $24 - 3x$ must be greater than 0.

$24-3x>0$

$24>3x$

$8>x \quad \text{or} \quad x<8$

The domain is $(-\infty, 8)$.

Pencil Problem #1

1a. Find the domain of $f(x)=3(x-4)$.

1b. Find the domain of $g(x)=\dfrac{\sqrt{x-2}}{x-5}$.

Objective #2: Combine functions using the algebra of functions, specifying domains.

✔ Solved Problem #2

2. Let $f(x) = x - 5$ and $g(x) = x^2 - 1$. Find each function and determine its domain.

2a. $(f+g)(x)$

$$\begin{aligned}(f+g)(x) &= f(x) + g(x)\\ &= (x-5) + (x^2-1)\\ &= x^2 + x - 6\end{aligned}$$

The domain is $(-\infty, \infty)$.

2b. $(f-g)(x)$

$$\begin{aligned}(f-g)(x) &= f(x) - g(x)\\ &= (x-5) - (x^2-1)\\ &= x - 5 - x^2 + 1\\ &= -x^2 + x - 4\end{aligned}$$

The domain is $(-\infty, \infty)$.

2c. $(fg)(x)$

$$\begin{aligned}(fg)(x) &= f(x) \cdot g(x)\\ &= (x-5)(x^2-1)\\ &= x^3 - 5x^2 - x + 5\end{aligned}$$

The domain is $(-\infty, \infty)$.

2d. $\left(\frac{f}{g}\right)(x)$

$$\left(\frac{f}{g}\right)(x) = \frac{f(x)}{g(x)} = \frac{x-5}{x^2-1}$$

The function contains division; it is undefined when $x^2 - 1 = 0$ or $x^2 = 1$ or $x = \pm 1$. The domain is $(-\infty, -1) \cup (-1, 1) \cup (1, \infty)$.

Pencil Problem #2

2. Let $f(x) = 2x^2 - x - 3$ and $g(x) = x + 1$. Find each function and determine its domain.

2a. $(f+g)(x)$

2b. $(f-g)(x)$

2c. $(fg)(x)$

2d. $\left(\frac{f}{g}\right)(x)$

Objective #3: Form composite functions.

✔ *Solved Problem #3*

3. Given $f(x) = 5x + 6$ and $g(x) = 2x^2 - x - 1$, find each of the following.

3a. $(f \circ g)(x)$

$$\begin{aligned}(f \circ g)(x) &= f(g(x))\\ &= 5g(x) + 6\\ &= 5(2x^2 - x - 1) + 6\\ &= 10x^2 - 5x - 5 + 6\\ &= 10x^2 - 5x + 1\end{aligned}$$

3b. $(g \circ f)(x)$

$$\begin{aligned}(g \circ f)(x) &= g(f(x))\\ &= 2[f(x)]^2 - f(x) - 1\\ &= 2(5x+6)^2 - (5x+6) - 1\\ &= 2(25x^2 + 60x + 36) - 5x - 6 - 1\\ &= 50x^2 + 120x + 72 - 5x - 7\\ &= 50x^2 + 115x + 65\end{aligned}$$

3c. $(f \circ g)(-1)$

$$\begin{aligned}(f \circ g)(-1) &= 10(-1)^2 - 5(-1) + 1\\ &= 10 + 5 + 1\\ &= 16\end{aligned}$$

Pencil Problem #3

3. Given $f(x) = 4x - 3$ and $g(x) = 5x^2 - 2$, find each of the following.

3a. $(f \circ g)(x)$

3b. $(g \circ f)(x)$

3c. $(f \circ g)(2)$

Objective #4: Determine domains for composite functions.

✔ *Solved Problem #4*

4. Given $f(x) = \dfrac{4}{x+2}$ and $g(x) = \dfrac{1}{x}$, find each of the following.

4a. $(f \circ g)(x)$

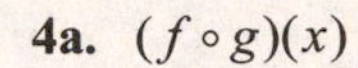

$$(f \circ g)(x) = \frac{4}{g(x)+2} = \frac{4}{\frac{1}{x}+2} \cdot \frac{x}{x} = \frac{4x}{1+2x}$$

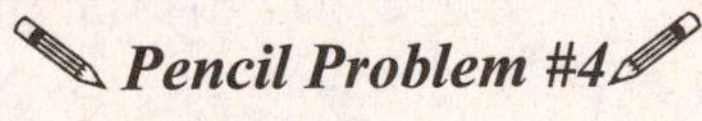

Pencil Problem #4

4. Given $f(x) = \dfrac{2}{x+3}$ and $g(x) = \dfrac{1}{x}$, find each of the following.

4a. $(f \circ g)(x)$

4b. The domain of $f \circ g$

The function g is undefined when $x = 0$, so 0 is not in the domain of $f \circ g$. The function f is undefined for $x = -2$, so any values of x for which $g(x) = -2$ are not in the domain of $f \circ g$. Solving $\frac{1}{x} = -2$, we find that $x = -\frac{1}{2}$.

The domain is $\left(-\infty, -\frac{1}{2}\right) \cup \left(-\frac{1}{2}, 0\right) \cup (0, \infty)$.

4b. The domain of $f \circ g$

Objective #5: Write functions as compositions.

✔ Solved Problem #5

5. Express $h(x) = \sqrt{x^2 + 5}$ as the composition of two functions.

A natural way to write h as the composition of two functions is to take the square root of $g(x) = x^2 + 5$.

Let $f(x) = \sqrt{x}$ and $g(x) = x^2 + 5$.

Then $(f \circ g)(x) = \sqrt{g(x)} = \sqrt{x^2 + 5} = h(x)$.

Pencil Problem #5

5. Express $h(x) = (3x - 1)^4$ as the composition of two functions.

Answers for Pencil Problems *(Textbook Exercise references in parentheses)*:

1a. $(-\infty, \infty)$ *(2.6 #1)* **1b.** $[2, 5) \cup (5, \infty)$ *(2.6 #27)*

2a. $(f + g)(x) = 2x^2 - 2$; domain: $(-\infty, \infty)$ *(2.6 #35)*

2b. $(f - g)(x) = 2x^2 - 2x - 4$; domain: $(-\infty, \infty)$ *(2.6 #35)*

2c. $(fg)(x) = 2x^3 + x^2 - 4x - 3$; domain: $(-\infty, \infty)$ *(2.6 #35)*

2d. $\left(\frac{f}{g}\right)(x) = \frac{2x^2 - x - 3}{x + 1}$; domain: $(-\infty, -1) \cup (1, \infty)$ *(2.6 #35)*

3a. $(f \circ g)(x) = 20x^2 - 11$ *(2.6 #55a)* **3b.** $(g \circ f)(x) = 80x^2 - 120x + 43$ *(2.6 #55 b)* **3c.** 69 *(2.6 #55c)*

4a. $(f \circ g)(x) = \frac{2x}{1 + 3x}$ *(2.6 #67a)* **4b.** $\left(-\infty, -\frac{1}{3}\right) \cup \left(-\frac{1}{3}, 0\right) \cup (0, \infty)$ *(2.6 #67b)*

5. Let $f(x) = x^4$ and $g(x) = 3x - 1$. Then $(f \circ g)(x) = h(x)$. *(2.6 #75)*

Section 2.7
Inverse Functions

Hey! That's My Birthday Too !

What is the probability that two people in the same room share a birthday?

It might be higher than you think.

In this section we will explore the graph of the function that represents this probability.

Objective #1: Verify inverse functions.

✔ *Solved Problem #1*

1. Show that each function is the inverse of the other:

$f(x)=4x-7$ and $g(x)=\dfrac{x+7}{4}$.

First, show that $f(g(x))=x$.

$$f(g(x))=4\left(\frac{x+7}{4}\right)-7$$
$$=x+7-7$$
$$=x$$

Next, show that $g(f(x))=x$.

$$g(f(x))=\frac{(4x-7)+7}{4}$$
$$=\frac{4x}{4}$$
$$=x$$

Pencil Problem #1

1. Determine whether $f(x)=\dfrac{3}{x-4}$ and $g(x)=\dfrac{3}{x}+4$ are inverses of each other.

Objective #2: Find the inverse of a function.

✔ *Solved Problem #2*

2a. Find the inverse of $f(x)=2x+7$.

Replace $f(x)$ with y.

$y=2x+7$

Interchange x and y and solve for y.

$$x=2y+7$$
$$x-7=2y$$
$$\frac{x-7}{2}=y$$

Replace y with $f^{-1}(x)$.

$$f^{-1}(x)=\frac{x-7}{2}$$

Pencil Problem #2

2a. Find the inverse of $f(x)=x+3$.

2b. Find the inverse of $f(x)=4x^3-1$.

Replace $f(x)$ with y.

$$y=4x^3-1$$

Interchange x and y and solve for y.

$$x=4y^3-1$$
$$x+1=4y^3$$
$$\frac{x+1}{4}=y^3$$
$$\sqrt[3]{\frac{x+1}{4}}=y$$

Replace y with $f^{-1}(x)$.

$$f^{-1}(x)=\sqrt[3]{\frac{x+1}{4}}$$

2b. Find the inverse of $f(x)=(x+2)^3$.

2c. Find the inverse of $f(x)=\frac{3}{x}-1$.

Replace $f(x)$ with y.

$$y=\frac{3}{x}-1$$

Interchange x and y and solve for y.

$$x=\frac{3}{y}-1$$
$$xy=\left(\frac{3}{y}-1\right)y$$
$$xy=\frac{3}{y}\cdot y-1\cdot y$$
$$xy=3-y$$
$$xy+y=3$$
$$y(x+1)=3$$
$$\frac{y(x+1)}{x+1}=\frac{3}{x+1}$$
$$y=\frac{3}{x+1}$$

Replace y with $f^{-1}(x)$.

$$f^{-1}(x)=\frac{3}{x+1}$$

2c. Find the inverse of $f(x)=\frac{7}{x}-3$.

Objective #3: Use the horizontal line test to determine if a function has an inverse function.

✔ *Solved Problem #3*

3. Use the horizontal line test to determine if the following graph represents a function that has an inverse function.

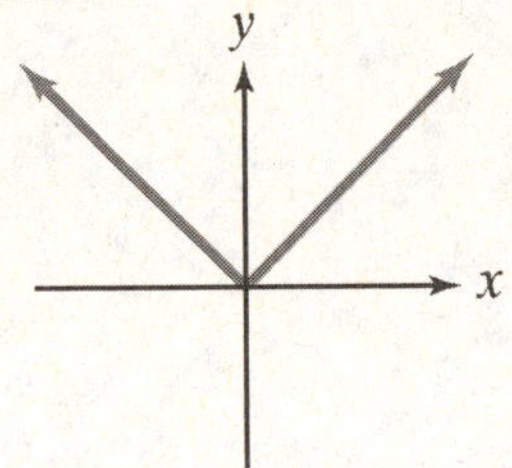

Since a horizontal line can be drawn that intersects the graph more than once, it fails the horizontal line test.

Thus, this graph does not represent a function that has an inverse function.

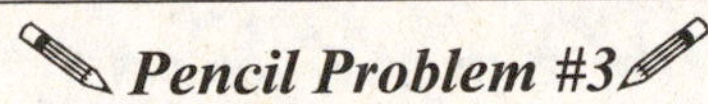

Pencil Problem #3

3. Use the horizontal line test to determine if the following graph represents a function that has an inverse function.

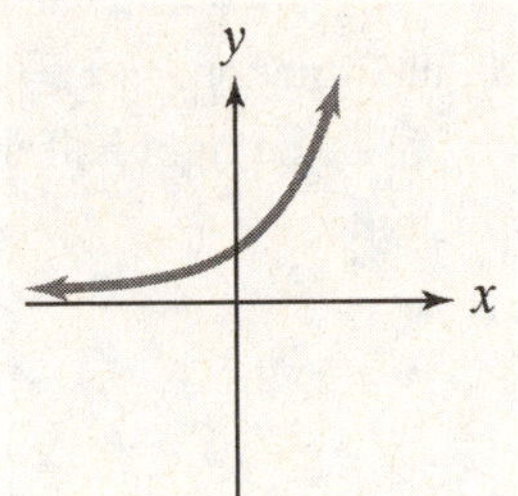

Objective #4: Use the graph of a one-to-one function to graph its inverse function.

✔ *Solved Problem #4*

4. The graph of function f consists of two line segments, one segment from $(-2,-2)$ to $(-1,0)$, and a second segment from $(-1,0)$ to $(1,2)$. Graph f and use the graph to draw the graph of its inverse function.

Since f has a line segment from $(-2,-2)$ to $(-1,0)$, then f^{-1} has a line segment from $(-2,-2)$ to $(0,-1)$.

Since f has a line segment from $(-1,0)$ to $(1,2)$, then f^{-1} has a line segment from $(0,-1)$ to $(2,1)$.

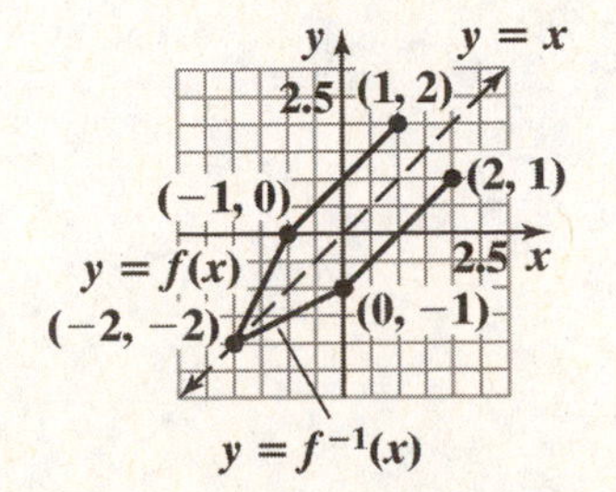

Pencil Problem #4

4. The graph of a linear function f contains the points $(0,-4)$, $(2,0)$, $(3,2)$, and $(4,4)$. Draw the graph of the inverse function.

Objective #5: Find the inverse of a function and graph both functions on the same axes.

✔ Solved Problem #5

5. Find the inverse of $f(x) = x^2 + 1$ if $x \geq 0$. Graph f and f^{-1} in the same rectangular coordinate system.

Restricted to $x \geq 0$, the function $f(x) = x^2 + 1$ has an inverse. The graph of f is the right half of the graph of $y = x^2$ shifted up 1 unit.

Replace $f(x)$ with y: $y = x^2 + 1$.

Interchange x and y and solve for y. Since the values of x are nonnegative in the original function, the values of y must be nonnegative in the inverse function. We choose the positive square root in the third step below.

$$x = y^2 + 1$$
$$x - 1 = y^2$$
$$\sqrt{x-1} = y$$

Replace y with $f^{-1}(x)$: $f^{-1}(x) = \sqrt{x-1}$.

The graph of f^{-1} is the graph of the square root function shifted 1 unit to the left. The graph of f^{-1} is also the reflection of the graph of f about the line $y = x$.

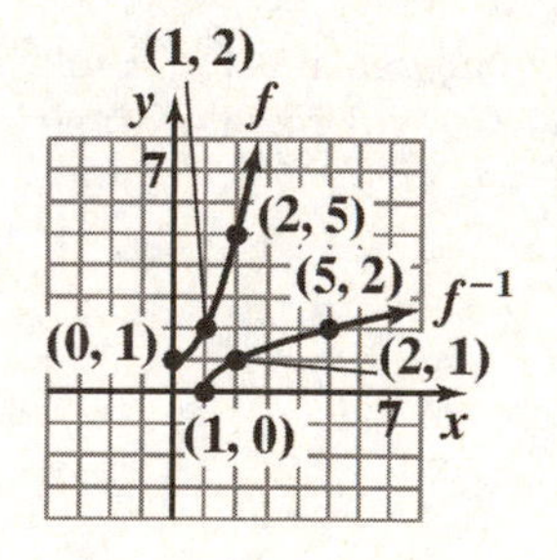

Pencil Problem #5

5. Find the inverse of $f(x) = (x-1)^2$ if $x \leq 1$. Graph f and f^{-1} in the same rectangular coordinate system.

Answers for Pencil Problems *(Textbook Exercise references in parentheses)*:

1. The functions are inverses of each other. *(2.7 #7)* **2a.** $f^{-1}(x) = x - 3$ *(2.7 #11)*

2b. $f^{-1}(x) = \sqrt[3]{x} - 2$ *(2.7 #19)* **2c.** $f^{-1}(x) = \dfrac{7}{x+3}$ *(2.7 #25)* **3.** has inverse function *(2.7 #33)*

4.

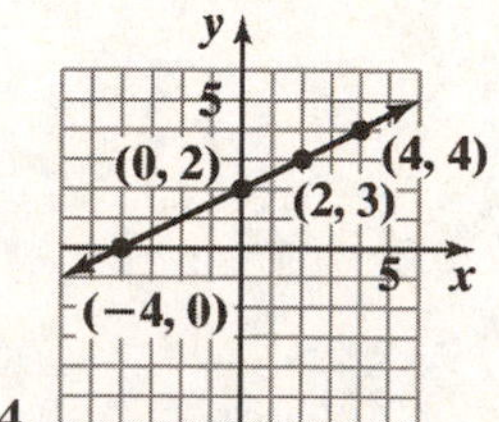

(2.7 #35)

5.

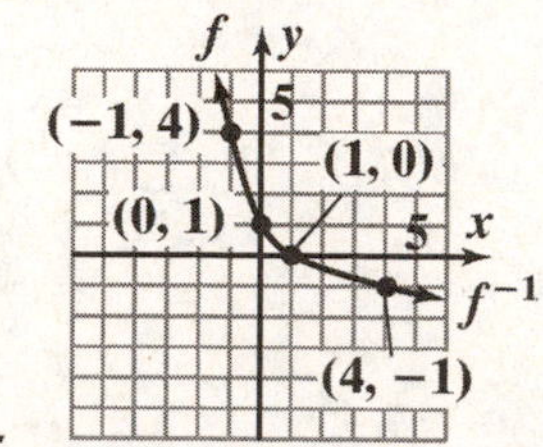

(2.7 #43)

Section 2.8
Distance and Midpoint Formulas; Circles

Round and Round !

In 1893, George Washington Gale Ferris, Jr. designed and built the first Ferris wheel as the centerpiece for the World's Columbian Exposition in Chicago.

The rectangular coordinate system gives us a unique way of knowing a circle. It enables us to translate a circle's geometric definition into an algebraic equation. In this section, we will learn, and then apply, these algebraic techniques.

Objective #1: Find the distance between two points.

✔ *Solved Problem #1*

1. Find the distance between $(-1,-3)$ and $(2,3)$.
Express the answer in simplified radical form and then round to two decimal places.

$$d=\sqrt{(x_2-x_1)^2+(y_2-y_1)^2}$$
$$d=\sqrt{(2-(-1))^2+(3-(-3))^2}$$
$$=\sqrt{3^2+6^2}$$
$$=\sqrt{9+36}$$
$$=\sqrt{45}$$
$$=3\sqrt{5}$$
$$\approx 6.71 \text{ units}$$

Pencil Problem #1

1. Find the distance between $(4,1)$ and $(6,3)$.
Express the answer in simplified radical form and then round to two decimal places.

Objective #2: Find the midpoint of a line segment.

✔ *Solved Problem #2*

2. Find the midpoint of the line segment with endpoints $(1,2)$ and $(7,-3)$.

$$\text{Midpoint}=\left(\frac{x_1+x_2}{2},\frac{y_1+y_2}{2}\right)$$
$$=\left(\frac{1+7}{2},\frac{2+(-3)}{2}\right)$$
$$=\left(\frac{8}{2},\frac{-1}{2}\right)$$
$$=\left(4,-\frac{1}{2}\right)$$

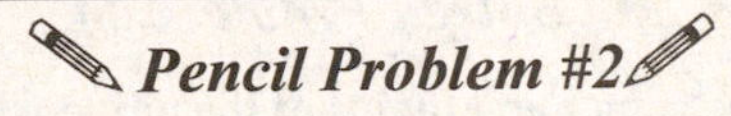

Pencil Problem #2

2. Find the midpoint of the line segment with endpoints $(6,8)$ and $(2,4)$.

Objective #3: Write the standard form of a circle's equation.

✔ Solved Problem #3

3a. Write the standard form of the equation of the circle with center $(0,0)$ and radius 4.

$$(x-h)^2+(y-k)^2=r^2$$
$$(x-0)^2+(y-0)^2=4^2$$
$$x^2+y^2=16$$

Pencil Problem #3

3a. Write the standard form of the equation of the circle with center $(0,0)$ and radius 7.

✔ Solved Problem #3

3b. Write the standard form of the equation of the circle with center $(0,-6)$ and radius 10.

$$(x-h)^2+(y-k)^2=r^2$$
$$(x-0)^2+(y-(-6))^2=10^2$$
$$x^2+(y+6)^2=100$$

Pencil Problem #3

3b. Write the standard form of the equation of the circle with center $(-1,4)$ and radius 2.

Objective #4: Give the center and radius of a circle whose equation is in standard form.

✔ Solved Problem #4

4a. Find the center and radius of the circle whose equation is $(x+3)^2+(y-1)^2=4$.

$$(x+3)^2+(y-1)^2=4$$
$$(x-(-3))^2+(y-1)^2=2^2$$

The center is $(-3,1)$ and the radius is 2 units.

4a. Find the center and radius of the circle whose equation is $(x-3)^2+(y-1)^2=36$.

4b. Graph the equation in Solved Problem 4a.

Plot points 2 units above and below and to the left and right of the center, (–3,1). Draw a circle through these points.

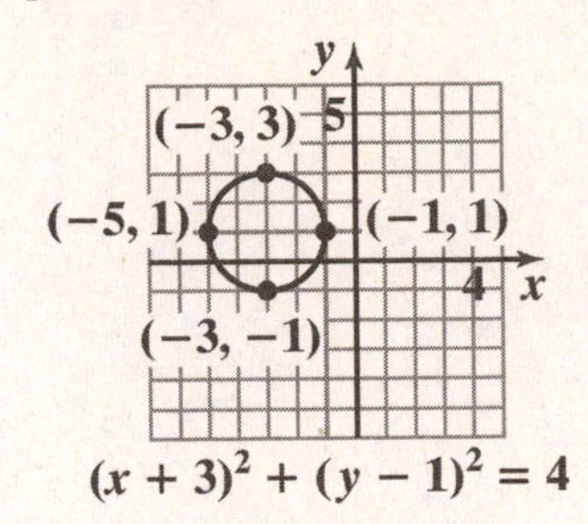

4b. Graph the equation in Pencil Problem 4a.

4c. Use the graph in Solved Problem 4b to identify the relation's domain and range.

The leftmost point on the circle has an x-coordinate of –5, and the rightmost point has an x-coordinate of –1. The domain is [–5, –1].

The lowest point on the graph has a y-coordinate of –1, and the highest point on the graph has a y-coordinate of 3. The range is [–1, 3].

4c. Use the graph in Pencil Problem 4b to identify the relation's domain and range.

Objective #5: Convert the general form of a circle's equation to standard form.

✔ *Solved Problem #5*

5. Write in standard form and graph:

$$x^2 + y^2 + 4x - 4y - 1 = 0$$

$$x^2 + y^2 + 4x - 4y - 1 = 0$$

$$\left(x^2 + 4x \quad\right) + \left(y^2 - 4y \quad\right) = 1$$

Complete the squares.

For x: $\left(\frac{b}{2}\right)^2 = \left(\frac{4}{2}\right)^2 = (2)^2 = 4$

For y: $\left(\frac{b}{2}\right)^2 = \left(\frac{-4}{2}\right)^2 = (-2)^2 = 4$

(Continued on next page)

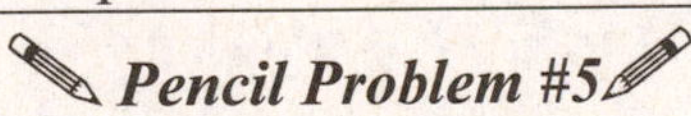

Pencil Problem #5

5. Write in standard form and graph:

$$x^2 + y^2 + 8x - 2y - 8 = 0$$

Add these values to both sides of the equation.

$$x^2+y^2+4x-4y-1=0$$
$$\left(x^2+4x\quad\right)+\left(y^2-4y\quad\right)=1$$
$$\left(x^2+4x+4\right)+\left(y^2-4y+4\right)=1+4+4$$
$$(x+2)^2+(y-2)^2=9$$
$$(x-(-2))^2+(y-2)^2=3^2$$

The center is $(-2,2)$ and the radius is 3 units.

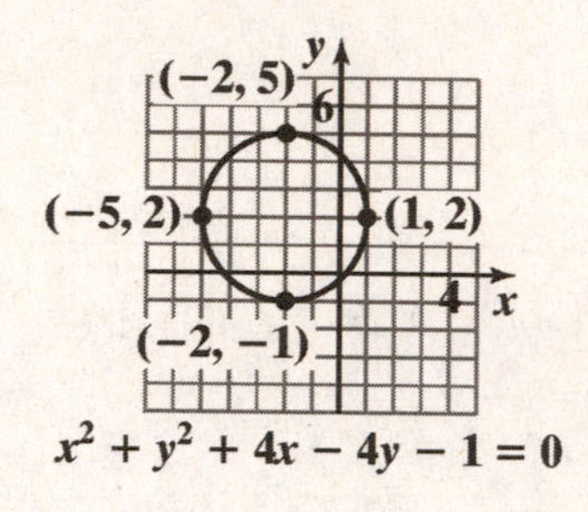

$x^2 + y^2 + 4x - 4y - 1 = 0$

Answers for Pencil Problems *(Textbook Exercise references in parentheses)*:

1. $2\sqrt{2} \approx 2.83$ *(2.8 #3)* **2.** $(4,6)$ *(2.8 #19)*

3a. $x^2+y^2=49$ *(2.8 #31)*

3b. $(x+1)^2+(y-4)^2=4$ *(2.8 #35)*

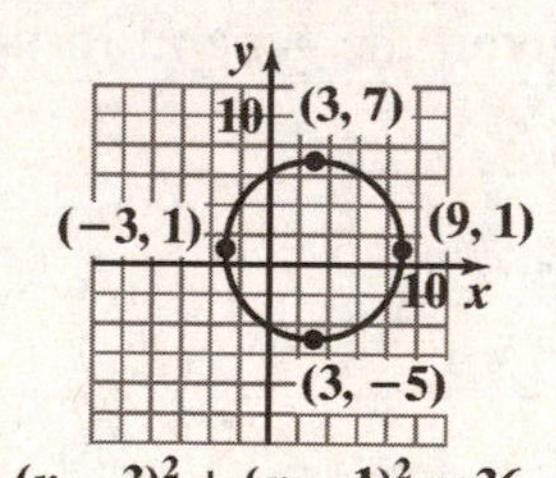

4a. center: $(3,1)$; radius: 6 units

4b. $(x-3)^2+(y-1)^2=36$ *(2.8 #43)*

4c. domain: $[-3,9]$; range: $[-5,7]$ *(2.8 #43)*

5. $(x+4)^2+(y-1)^2=25$;

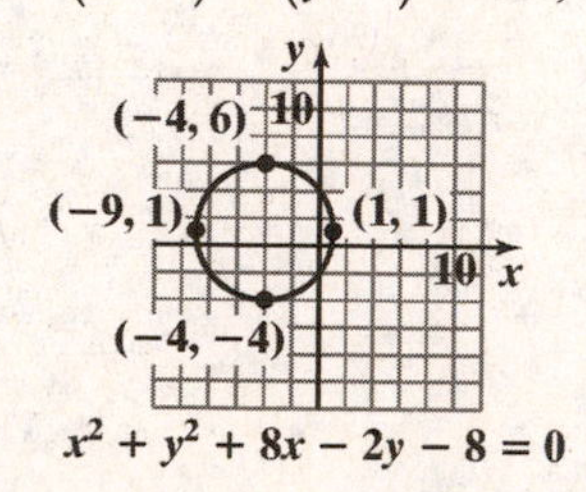

$x^2 + y^2 + 8x - 2y - 8 = 0$ *(2.8 #57)*

Section 3.1
Quadratic Functions and Their Graphs

Heads UP!!!

Many sports involve objects that are thrown, kicked, or hit, and then proceed with no additional force of their own.
Such objects are called projectiles.

In this section of your textbook, you will learn to use graphs of quadratic functions to gain a visual understanding of various projectile sports.

Objective #1: Recognize characteristics of parabolas.

✔ *Solved Problem #1*

1. True or false: The *vertex* of a parabola is also called the *turning point*.

True

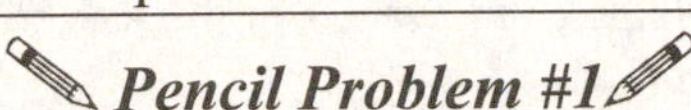

Pencil Problem #1

1. True or false: The *vertex* of a parabola is always the minimum point of the parabola.

Objective #2: Graph parabolas.

✔ *Solved Problem #2*

2a. Graph the quadratic function: $f(x) = -(x-1)^2 + 4$

Since $a = -1$ is negative, the parabola opens downward.

The vertex of the parabola is $(h,k) = (1,4)$.

Replace $f(x)$ with 0 to find x–intercepts.

$$0 = -(x-1)^2 + 4$$
$$(x-1)^2 = 4$$
$$x-1 = \pm\sqrt{4}$$
$$x-1 = \pm 2$$
$$x-1 = 2 \quad \text{or} \quad x-1 = -2$$
$$x = 3 \qquad\qquad x = -1$$

The x–intercepts are -1 and 3.
Set $x = 0$ and solve for y to obtain the y–intercept.

(Continued on next page)

Pencil Problem #2

2a. Graph the quadratic function: $f(x) = (x-4)^2 - 1$

$y = -(0-1)^2 + 4 = 3$

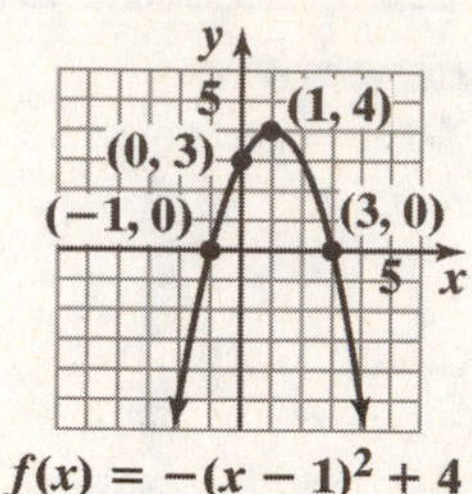

2b. Graph the quadratic function $f(x) = -x^2 + 4x + 1$. Use the graph to identify the function's domain and its range.

Since $a = -1$ is negative, the parabola opens downward.

The x–coordinate of the vertex of the parabola is
$-\frac{b}{2a} = -\frac{4}{2(-1)} = -\frac{4}{-2} = 2.$

The y–coordinate of the vertex of the parabola is
$f\left(-\frac{b}{2a}\right) = f(2) = -(2)^2 + 4(2) + 1 = 5.$

The vertex is $(2,5)$.

Replace $f(x)$ with 0 to find x–intercepts.

$0 = -x^2 + 4x + 1$

$x = \frac{-b \pm \sqrt{b^2 - 4ac}}{2a}$

$x = \frac{-4 \pm \sqrt{4^2 - 4(-1)(1)}}{2(-1)}$

$x = 2 \pm \sqrt{5}$

$x \approx -0.2$ or $x \approx 4.2$

The x–intercepts are -0.2 and 4.2.

Set $x = 0$ and solve for y to obtain the y–intercept.

$y = -0^2 + 4 \cdot 0 + 1 = 1$

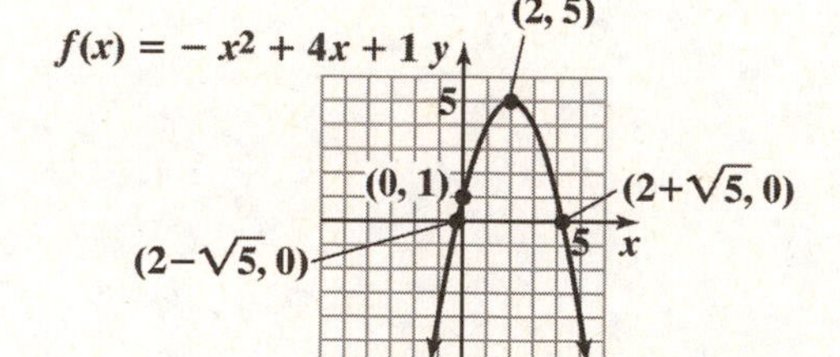

Domain: $(-\infty, \infty)$ Range: $(-\infty, 5]$

2b. Graph the quadratic function $f(x) = x^2 + 3x - 10$. Use the graph to identify the function's range.

Objective #3: Determine a quadratic function's minimum or maximum value.

✔ Solved Problem #3

3. Consider the quadratic function
$f(x) = 4x^2 - 16x + 1000$.

3a. Determine, without graphing, whether the function has a minimum value or a maximum value.

Because $a > 0$, the function has a minimum value.

3b. Find the minimum or maximum value and determine where it occurs.

The minimum value occurs at $-\frac{b}{2a} = -\frac{-16}{2(4)} = 2$.

The minimum of $f(x)$ is $f(2) = 4 \cdot 2^2 - 16 \cdot 2 + 1000$
$= 984$.

3c. Identify the function's domain and its range.

Like all quadratic functions, the domain is $(-\infty, \infty)$.

Because the minimum is 984, the range includes all real numbers at or above 984. The range is $[984, \infty)$.

Pencil Problem #3

3. Consider the quadratic function
$f(x) = -4x^2 + 8x - 3$.

3a. Determine, without graphing, whether the function has a minimum value or a maximum value.

3b. Find the minimum or maximum value and determine where it occurs.

3c. Identify the function's domain and its range.

Objective #4: Solve problems involving a quadratic function's minimum or maximum value.

✔ Solved Problem #4

4. Among all pairs of numbers whose difference is 8, find a pair whose product is as small as possible. What is the minimum product?

Let the two numbers be represented by x and y, and let the product be represented by P.

We must minimize $P = xy$.

Because the difference of the two numbers is 8, then $x - y = 8$.

Solve for y in terms of x.

$x - y = 8$
$-y = -x + 8$
$y = x - 8$

(Continued on next page)

Pencil Problem #4

4. Among all pairs of numbers whose sum is 16, find a pair whose product is as large as possible. What is the maximum product?

Write P as a function of x.

$$P = xy$$
$$P(x) = x(x-8)$$
$$P(x) = x^2 - 8x$$

Because $a > 0$, the function has a minimum value that occurs at $x = -\frac{b}{2a}$

$$= -\frac{-8}{2(1)}$$
$$= 4.$$

Substitute to find the other number.

$$y = x - 8$$
$$y = 4 - 8$$
$$= -4$$

The two numbers are 4 and −4.

The minimum product is $P = xy = (4)(-4) = -16$.

Answers for Pencil Problems *(Textbook Exercise references in parentheses)*:

1. false *(3.1 #41)*

2a.

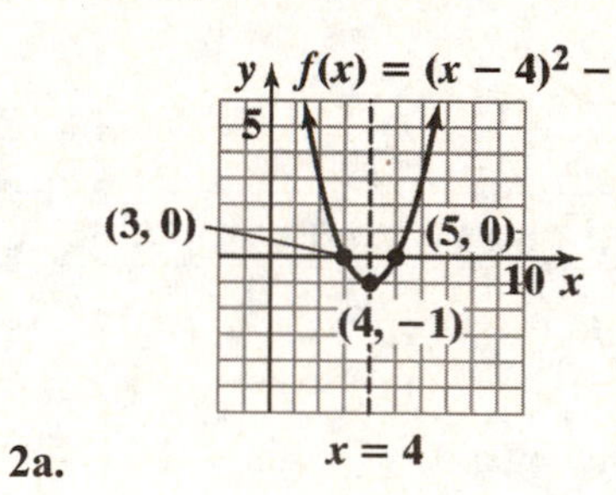

(3.1 #17)

2b.

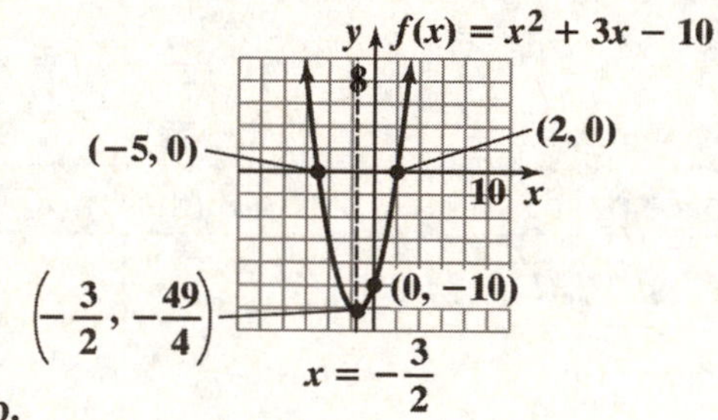

Range: $\left[-\frac{49}{4}, \infty\right)$ *(3.1 #29)*

3a. maximum *(3.1 #41a)*

3b. The maximum is 1 at $x = 1$. *(3.1 #41b)*

3c. Domain: $(-\infty, \infty)$; Range: $(-\infty, 1]$ *(3.1 #41c)*

4. The maximum product is 64 when the numbers are 8 and 8. *(3.1 #61)*

Section 3.2
Polynomial Functions and Their Graphs

Pay at the Pump !

Other than outrage, what is going on at the gas pumps?
Is surging demand creating the increasing oil prices?
Like all things in a free market economy, the price of a commodity is based on supply and demand.

In the Exercise Set for this section, we will explore the volatility of gas prices over the past several years.

Objective #1: Identify polynomial functions.

✔ *Solved Problem #1*

1. The exponents on the variables in a polynomial function must be nonnegative integers.

True

Pencil Problem #1

1. The coefficients of the variables in a polynomial function must be nonnegative integers.

Objective #2: Recognize characteristics of graphs of polynomial functions.

✔ *Solved Problem #2*

2. The graph of a polynomial function may have a sharp corner.

False. The graphs of polynomial functions are smooth, meaning that they have rounded curves and no sharp corners.

Pencil Problem #2

2. The graph of a polynomial function may have a gap or break.

Objective #3: Determine end behavior.

✔ *Solved Problem #3*

3. Use the Leading Coefficient Test to determine the end behavior of the graph of each function.

3a. $f(x) = x^4 - 4x^2$

The term with the greater exponent is x^4, or $1x^4$. The leading coefficient is 1, which is positive. The degree of the function is 4, which is even. Even-degree polynomial functions have the same behavior at each end. Since the leading coefficient is positive, the graph rises to the left and rises to the right.

3. Use the Leading Coefficient Test to determine the end behavior of the graph of each function.

3a. $f(x) = 5x^3 + 7x^2 - x + 9$

3b. $f(x) = 2x^3(x-1)(x+5)$

The function is in factored form, but we can determine the degree and the leading coefficient without multiplying it out. The factors $2x^3$, $x - 1$, and $x + 5$ are of degree 3, 1, and 1, respectively. When we multiply expressions with the same base, we add exponents, so the degree of the function is 3 + 1 + 1, or 5, which is odd. Without multiplying out, you should be able to see that the leading coefficient is 2, which is positive.

Odd-degree polynomial functions have graphs with opposite behavior at each end. Since the leading coefficient is positive, the graph falls to the left and rises to the right.

3b. $f(x) = -x^2(x-1)(x+3)$

Objective #4: Use factoring to find zeros of polynomial functions.

✔ *Solved Problem #4*

4. Find all zeros of $f(x) = x^3 + 2x^2 - 4x - 8$.

Set $f(x)$ equal to zero.

$$x^3 + 2x^2 - 4x - 8 = 0$$
$$x^2(x+2) - 4(x+2) = 0$$
$$(x+2)(x^2 - 4) = 0$$
$$(x+2)(x+2)(x-2) = 0$$

Apply the zero-product principle.

$x + 2 = 0$ or $x + 2 = 0$ or $x - 2 = 0$

$x = -2$ $\quad$ $x = -2$ $\quad$ $x = 2$

The zeros are -2 and 2.

Pencil Problem #4

4. Find all zeros of $f(x) = x^3 + 2x^2 - x - 2$.

Objective #5: Identify zeros and their multiplicities.

✔ *Solved Problem #5*

5. Find the zeros of $f(x) = -4\left(x + \frac{1}{2}\right)^2 (x-5)^3$ and give the multiplicity of each zero. State whether the graph crosses the x-axis or touches the x-axis and turns around at each zero.

Set each factor equal to zero.

$x + \frac{1}{2} = 0$ or $x - 5 = 0$

$x = -\frac{1}{2}$ $\quad$ $x = 5$

$-\frac{1}{2}$ is a zero of multiplicity 2, and 5 is a zero of multiplicity 3.

Because the multiplicity of $-\frac{1}{2}$ is even, the graph touches the x-axis and turns around at this zero.

Because the multiplicity of 5 is odd, the graph crosses the x-axis at this zero.

Pencil Problem #5

5. Find the zeros of $f(x) = 4(x-3)(x+6)^3$ and give the multiplicity of each zero. State whether the graph crosses the x-axis or touches the x-axis and turns around at each zero.

Objective #6: Use the Intermediate Value Theorem.

✔ *Solved Problem #6*

6. Show that the polynomial function $f(x) = 3x^3 - 10x + 9$ has a real zero between -3 and -2.

Evaluate f at -3 and -2.

$f(-3) = 3(-3)^3 - 10(-3) + 9 = -42$

$f(-2) = 3(-2)^3 - 10(-2) + 9 = 5$

The sign change between $f(-3)$ and $f(-2)$ shows that f has a real zero between -3 and -2.

Pencil Problem #6

6. Show that the polynomial function $f(x) = 2x^4 - 4x^2 + 1$ has a real zero between -1 and 0.

Objective #7: Understand the relationship between degree and turning points.

✔ *Solved Problem #7*

7. If a polynomial function, f, is of degree 5, what is the greatest number of turning points on its graph?

The greatest number of turning points on the graph of a polynomial of degree 5 is $5-1$, or 4.

Pencil Problem #7

7. If a polynomial function, f, is of degree 4, what is the greatest number of turning points on its graph?

Objective #8: Graph polynomial functions.

✔ *Solved Problem #8*

8. Use the five-step strategy to graph $f(x) = x^3 - 3x^2$.

Step 1: Determine end behavior.

Since $f(x) = x^3 - 3x^2$ is an odd-degree polynomial and since the leading coefficient, 1, is positive, the graph falls to the left and rises to the right.

Step 2: Find x-intercepts by setting $f(x) = 0$.

$x^3 - 3x^2 = 0$

$x^2(x-3) = 0$

Apply the zero-product principle.

$x^2 = 0$ or $x - 3 = 0$

$x = 0$ $\quad\quad$ $x = 3$

The zeros of f are 0 and 3. The graph touches the x-axis at 0 since it has multiplicity 2. The graph crosses the x-axis at 3 since it has multiplicity 1.

(Continued on next page)

Pencil Problem #8

8. Use the five-step strategy to graph $f(x) = x^4 - 9x^2$.

Step 3: Find the y-intercept by computing $f(0)$.

$$f(x) = x^3 - 3x^2$$
$$f(0) = 0^3 - 3(0)^2$$
$$= 0$$

There is a y-intercept at 0, so the graph passes through (0, 0).

Step 4: Use possible symmetry to help draw the graph.

$$f(x) = x^3 - 3x^2$$
$$f(-x) = (-x)^3 - 3(-x)^2$$
$$= -x^3 - 3x^2$$

Since $f(-x) \neq f(x)$ and since $f(-x) \neq -f(x)$, the function is neither even nor odd, and the graph is neither symmetric with respect to the y-axis nor the origin.

Step 5: Draw the graph.

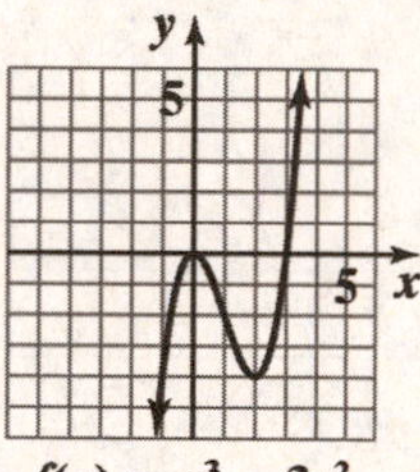

$f(x) = x^3 - 3x^2$

Answers for Pencil Problems *(Textbook Exercise references in parentheses)*:

1. False *(3.2 #3)* **2.** False *(3.2 #13)*

3a. The graph falls to the left and rises to the right. *(3.2 #19)*

3b. The graph falls to the left and falls to the right. *(3.2 #59a)*

4. -2, -1, and 1 *(3.2 #41b)*

5. zeros: 3 (multiplicity 1) and -6 (multiplicity 3); The graph crosses the x-axis at 3 and at -6. *(3.2 #27)*

6. $f(-1) = -1$ and $f(0) = 1$; The sign change between $f(-1)$ and $f(0)$ shows that f has a real zero between -1 and 0. *(3.2 #35)*

7. 3 *(3.2 #47e)*

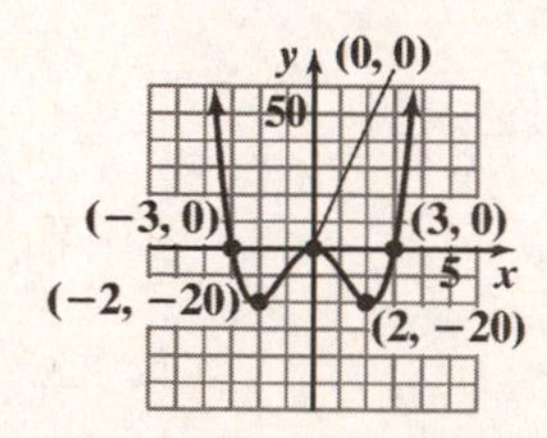

8. $f(x) = x^4 - 9x^2$ *(3.2 #43)*

Section 3.3
Dividing Polynomials; Remainder and Factor Theorems

What Happened to My Sweater?

It's that first brisk morning in autumn and you go to the closet for your favorite sweater. But what's that? There's a hole. No. There are dozens of holes.
In this section's Exercise Set, you will work with a polynomial function that models the number of eggs in a female moth based on her abdominal width. The techniques of this section provide a new way of evaluating the function to find out how many moths were eating your sweater.

Objective #1: Use long division to divide polynomials.

✔ *Solved Problem #1*

1. Divide $2x^4+3x^3-7x-10$ by x^2-2x.

Rewrite the dividend with the missing power of x and divide.

$$\begin{array}{r} 2x^2+7x+14 \\ x^2-2x\overline{\smash{)}2x^4+3x^3+0x^2-7x-10} \\ \underline{2x^4-4x^3} \\ 7x^3+0x^2 \\ \underline{7x^3-14x^2} \\ 14x^2-7x \\ \underline{14x^2-28x} \\ 21x-10 \end{array}$$

Thus, $\dfrac{2x^4+3x^3-7x-10}{x^2-2x}=2x^2+7x+14+\dfrac{21x-10}{x^2-2x}$

Pencil Problem #1

1. Divide $4x^4-4x^2+6x$ by $x-4$ using long division.

Objective #2: Use synthetic division to divide polynomials.

✔ *Solved Problem #2*

2. Use synthetic division: $(x^3-7x-6)\div(x+2)$

$$\begin{array}{r|rrrr} -2 & 1 & 0 & -7 & -6 \\ & & -2 & 4 & 6 \\ \hline & 1 & -2 & -3 & 0 \end{array}$$

Thus, $(x^3-7x-6)\div(x+2)=x^2-2x-3$

Pencil Problem #2

2. Use synthetic division: $(3x^2+7x-20)\div(x+5)$

Objective #3: Evaluate a polynomial function using the Remainder Theorem.

✔ Solved Problem #3

3. Given $f(x)=3x^3+4x^2-5x+3$, use the Remainder Theorem to find $f(-4)$.

$$\begin{array}{r|rrrr} -4 & 3 & 4 & -5 & 3 \\ & & -12 & 32 & -108 \\ \hline & 3 & -8 & 27 & -105 \end{array} \leftarrow f(-4)=-105$$

Pencil Problem #3

3. Given $f(x)=2x^3-11x^2+7x-5$, use the Remainder Theorem to find $f(4)$.

Objective #4: Use the Factor Theorem to solve a polynomial equation.

✔ Solved Problem #4

4. Solve the equation $15x^3+14x^2-3x-2=0$ given that -1 is a zero of $f(x)=15x^2+14x^2-3x-2$.

Synthetic division verifies that $x+1$ is a factor.

$$\begin{array}{r|rrrr} -1 & 15 & 14 & -3 & -2 \\ & & -15 & 1 & 2 \\ \hline & 15 & -1 & -2 & 0 \end{array}$$

Next, continue factoring to find all solutions.

$$15x^3+14x^2-3x-2=0$$
$$(x+1)(15x^2-x-2)=0$$
$$(x+1)(5x-2)(3x+1)=0$$
$$x+1=0 \quad \text{or} \quad 5x-2=0 \quad \text{or} \quad 3x+1=0$$
$$x=-1 \qquad 5x=2 \qquad 3x=-1$$
$$x=\frac{2}{5} \qquad x=-\frac{1}{3}$$

The solution set is $\left\{-1,-\frac{1}{3},\frac{2}{5}\right\}$

Pencil Problem #4

3 Solve the equation $2x^3-5x^2+x+2=0$ given that 2 is a zero of $f(x)=2x^3-5x^2+x+2$.

Answers for Pencil Problems *(Textbook Exercise references in parentheses)*:

1. $4x^3+16x^2+60x+246+\frac{984}{x-4}$ *(3.3 #11)*

2. $3x-8+\frac{20}{x+5}$ *(3.3 #19)*

3. -25 *(3.3 #33)*

4. $\left\{-\frac{1}{2}, 1, 2\right\}$ *(3.3 #43)*

Section 3.4
Zeros of Polynomials

Do I Have to Check My Bag?

Airlines have regulations on the sizes of carry-on luggage that are allowed. As a passenger, you are interested in the volume of your luggage, but the airline is concerned about the sum of bag's length, width, and depth.

In this section's Exercise Set, you will work with a polynomial function that relates the two quantities and allows you to find dimensions of a carry-on bag that meet both your volume requirement and the airline's regulations.

Objective #1: Use the Rational Zero Theorem to find possible rational zeros.

✔ *Solved Problem #1*

1. List all possible rational zeros of $f(x) = 4x^5 + 12x^4 - x - 3$.

Factors of the constant term -3: $\pm 1,\ \pm 3$

Factors of the leading coefficient 4: $\pm 1,\ \pm 2,\ \pm 4$

The possible rational zeros are:

$$\frac{\text{Factors of } -3}{\text{Factors of } 4} = \frac{\pm 1,\ \pm 3}{\pm 1,\ \pm 2,\ \pm 4}$$

$$= \pm 1,\ \pm 3,\ \pm\frac{1}{2},\ \pm\frac{3}{2},\ \pm\frac{1}{4},\ \pm\frac{3}{4}$$

Pencil Problem #1

1. List all possible rational zeros of $f(x) = 3x^4 - 11x^3 - x^2 + 19x + 6$.

Objective #2: Find zeros of a polynomial function.

✔ *Solved Problem #2*

2. Find all zeros of $f(x) = x^3 + x^2 - 5x - 2$.

First, list the possible rational zeros:

$$\frac{\text{Factors of } -2}{\text{Factors of } 1} = \frac{\pm 1,\ \pm 2}{\pm 1} = \pm 1,\ \pm 2$$

Now use synthetic division to find a rational zero from among the list of possible rational zeros. Try 2:

$$\begin{array}{r|rrrr} 2 & 1 & 1 & -5 & -2 \\ & & 2 & 6 & 2 \\ \hline & 1 & 3 & 1 & 0 \end{array}$$

The last number in the bottom row is 0.
Thus 2 is a zero and $x - 2$ is a factor.

The first three numbers in the bottom row of the synthetic division give the coefficients of the other factor. This factor is $x^2 + 3x + 1$.

(Continued on next page)

Pencil Problem #2

2. Find all zeros of $f(x) = x^3 + 4x^2 - 3x - 6$.

Factor completely: $x^3+x^2-5x-2=0$

$(x-2)(x^2+3x+1)=0$

Since x^2+3x+1 is not factorable, use the quadratic formula to find the remaining zeros.

$$x=\frac{-b\pm\sqrt{b^2-4ac}}{2a}$$

$$x=\frac{-3\pm\sqrt{3^2-4(1)(1)}}{2(1)}=\frac{-3\pm\sqrt{5}}{2}$$

The zeros are 2 and $\frac{-3\pm\sqrt{5}}{2}$.

Objective #3: Solve polynomial equations.

✔ *Solved Problem #3*

3. Solve: $x^4-6x^3+22x^2-30x+13=0$

First, list the possible rational roots:

$$\frac{\text{Factors of 13}}{\text{Factors of 1}}=\frac{\pm1,\ \pm13}{\pm1}=\pm1,\ \pm13$$

Now use synthetic division to find a rational root from among the list of possible rational roots. Try 1.

$$\begin{array}{r|rrrrr} 1 & 1 & -6 & 22 & -30 & 13 \\ & & 1 & -5 & 17 & -13 \\ \hline & 1 & -5 & 17 & -13 & 0 \end{array}$$

The last number in the bottom row is 0.
Thus, 1 is a root.

Rewrite the equation in factored form using the bottom row of the synthetic division to obtain the coefficients of the other factor.

$x^4-6x^3+22x^2-30x+13=0$

$(x-1)(x^3-5x^2+17x-13)=0$

Use the same approach to find another root. Try 1 again.

$$\begin{array}{r|rrrr} 1 & 1 & -5 & 17 & -13 \\ & & 1 & -4 & 13 \\ \hline & 1 & -4 & 13 & 0 \end{array}$$

The last number in the bottom row is 0.
Thus, 1 is a root (of multiplicity 2).

The first three numbers in the bottom row of the synthetic division give the coefficients of the factor $x^2-4x+13$.

(Continued on next page)

Pencil Problem #3

3. Solve: $x^3-2x^2-11x+12=0$

Since $x^2-4x+13$ is not factorable, use the quadratic formula to find the remaining roots.

$$x=\frac{-b\pm\sqrt{b^2-4ac}}{2a}$$
$$x=\frac{-(-4)\pm\sqrt{(-4)^2-4(1)(13)}}{2(1)}$$
$$x=\frac{4\pm\sqrt{-36}}{2}$$
$$x=\frac{4\pm 6i}{2}$$
$$x=2\pm 3i$$

The roots are 1 and $2\pm 3i$.

Objective #4: Use the Linear Factorization Theorem to find polynomials with given zeros.

✔ *Solved Problem #4*

4. Find a third-degree polynomial function $f(x)$ with real coefficients that has −3 and i as zeros such that $f(1)=8$.

Because i is a zero and the polynomial has real coefficients, the conjugate, $-i$, must also be a zero. We can now use the Linear Factorization Theorem.

$$\begin{aligned} f(x)&=a_n(x-c_1)(x-c_2)(x-c_3)\\ &=a_n(x-(-3))(x-i)(x-(-i))\\ &=a_n(x+3)(x-i)(x+i)\\ &=a_n(x+3)(x^2-i^2)\\ &=a_n(x+3)(x^2-(-1))\\ &=a_n(x+3)(x^2+1)\\ &=a_n(x^3+3x^2+x+3)\end{aligned}$$

Now we use $f(1)=8$ to find a_n.

$$f(1)=a_n(1^3+3\cdot 1^2+1+3)=8$$
$$8a_n=8$$
$$a_n=1$$

Now substitute 1 for a_n in the formula for $f(x)$.

$$f(x)=1(x^3+3x^2+x+3)$$
or $f(x)=x^3+3x^2+x+3$

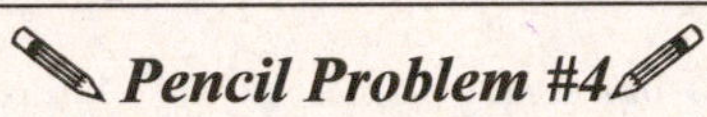

Pencil Problem #4

4. Find a fourth-degree polynomial function $f(x)$ with real coefficients that has i and $3i$ as zeros such that $f(-1)=20$.

Objective #5: Use Descartes's Rule of Signs.

✔ Solved Problem #5

5. Determine the possible numbers of positive and negative real zeros of $f(x) = x^4 - 14x^3 + 71x^2 - 154x + 120.$

Count the number of sign changes in $f(x)$.

$f(x) = x^4 - 14x^3 + 71x^2 - 154x + 120$

Since $f(x)$ has four sign changes, it has 4, 2, or 0 positive real zeros.

Count the number of sign changes in $f(-x)$.

$$f(-x) = (-x)^4 - 14(-x)^3 + 71(-x)^2 - 154(-x) + 120$$
$$= x^4 + 14x^3 + 71x^2 + 154x + 120$$

Since $f(-x)$ has no sign changes, $f(x)$ has 0 negative real zeros.

✎ Pencil Problem #5

5. Determine the possible numbers of positive and negative real zeros of $f(x) = x^3 + 2x^2 + 5x + 4.$

Answers for Pencil Problems *(Textbook Exercise references in parentheses)*:

1. $\pm 1,\ \pm 2,\ \pm 3,\ \pm 6,\ \pm\frac{1}{3},\ \pm\frac{2}{3}$ *(3.4 #3)*

2. $-1,\ \frac{-3-\sqrt{33}}{3}$, and $\frac{-3+\sqrt{33}}{3}$ *(3.4 #13)*

3. $\{-3, 1, 4\}$ *(3.4 #17)*

4. $f(x) = x^4 + 10x^2 + 9$ *(3.4 #29)*

5. f has no positive real zeros and either 3 or 1 negative real zeros *(3.4 #33)*

Section 3.5
Rational Functions and Their Graphs

Decreasing Costs with Increased Production?

In a simple business model, the cost, $C(x)$, to produce x units of a product is the sum of the fixed and variable costs and can be expressed in a form similar to $C(x) = \$500{,}000 + \$400x$. In this model, the cost increases by \$400 for each additional unit.

If we divide the cost, $C(x)$, by x, the number of units produced, we obtain the function $\bar{C}(x)$, which represents the average cost of each item. By studying the rational function $\bar{C}(x)$, we'll see that the average cost per item decreases for each additional unit.

Objective #1: Find the domains of rational functions.

✔ Solved Problem #1

1a. Find the domain of $g(x) = \dfrac{x}{x^2 - 25}$.

The denominator of $g(x) = \dfrac{x}{x^2 - 25}$ is 0 when $x = -5$ or $x = 5$. The domain of g consists of all real numbers except -5 and 5. This can be expressed as

$\{x \mid x \neq -5, x \neq 5\}$ or $(-\infty, -5) \cup (-5, 5) \cup (5, \infty)$

1b. Find the domain of $h(x) = \dfrac{x+5}{x^2 + 25}$.

No real numbers cause the denominator of $h(x) = \dfrac{x+5}{x^2 + 25}$ to equal 0. The domain of h consists of all real numbers, or $(-\infty, \infty)$.

Pencil Problem #1

1a. Find the domain of $h(x) = \dfrac{x+7}{x^2 - 49}$.

1b. Find the domain of $f(x) = \dfrac{x+7}{x^2 + 49}$.

Objective #2: Use arrow notation.

✔ Solved Problem #2

2. True or false: The notation "$x \to a^+$" means that the values of x are increasing without bound.

False. "$x \to a^+$" means that x is approaching a from the right.

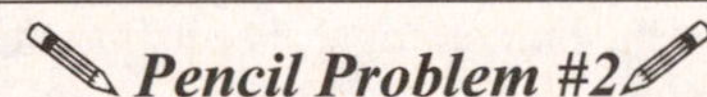

Pencil Problem #2

2. True or false: If $f(x) \to 0$ as $x \to \infty$, then the graph of f approaches the x-axis to the right.

Objective #3: Solve polynomial equations.

✔ *Solved Problem #3*

1a. Find the vertical asymptotes, if any, of the graph of the rational function: $g(x)=\dfrac{x-1}{x^2-1}$.

The numerator and denominator have a factor in common. Therefore, simplify g.

$$g(x)=\frac{x-1}{x^2-1}=\frac{x-1}{(x+1)(x-1)}=\frac{1}{x+1}$$

The only zero of the denominator of the simplified function is -1.

Thus, the line $x=-1$ is a vertical asymptote for the graph of g.

3b. Find the vertical asymptotes, if any, of the graph of the rational function: $h(x)=\dfrac{x-1}{x^2+1}$.

The denominator cannot be factored.
The denominator has no real zeros.

Thus, the graph of h has no vertical asymptotes.

Pencil Problem #3

3a. Find the vertical asymptotes, if any, of the graph of the rational function: $h(x)=\dfrac{x}{x(x+4)}$.

3b. Find the vertical asymptotes, if any, of the graph of the rational function: $r(x)=\dfrac{x}{x^2+4}$.

Objective #4: Identify horizontal asymptotes.

✔ *Solved Problem #4*

4a. Find the horizontal asymptotes, if any, of the graph of the rational function: $f(x)=\dfrac{9x^2}{3x^2+1}$.

The degree of the numerator, 2, is equal to the degree of the denominator, 2.
The leading coefficients of the numerator and denominator are 9 and 3, respectively.

Thus, the equation of the horizontal asymptote is $y=\dfrac{9}{3}$ or $y=3$.

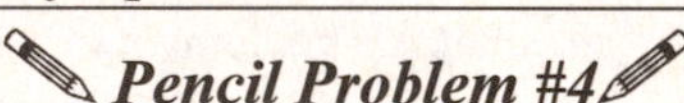

Pencil Problem #4

4a. Find the horizontal asymptotes, if any, of the graph of the rational function: $f(x)=\dfrac{-2x+1}{3x+5}$.

4b. Find the horizontal asymptotes, if any, of the graph of the rational function: $h(x)=\dfrac{9x^3}{3x^2+1}$.

The degree of the numerator, 3, is greater than the degree of the denominator, 2.

Thus, the graph of h has no horizontal asymptote.

4b. Find the horizontal asymptotes, if any, of the graph of the rational function: $f(x)=\dfrac{12x}{3x^2+1}$.

Objective #5: Use transformations to graph rational functions.

✔ *Solved Problem #5*

5. Use the graph of $f(x)=\dfrac{1}{x}$ to graph

$g(x)=\dfrac{1}{x+2}-1.$

Start with the graph of $f(x)=\dfrac{1}{x}$ and two points on its graph, such as $(-1,-1)$ and $(1, 1)$.

First move the graph two units to the left to graph $y=\dfrac{1}{x+2}$; the indicated points end up at $(-3,-1)$ and $(-1, 1)$. The vertical asymptote is now $x=-2$.

Next move the graph down one unit to graph $g(x)=\dfrac{1}{x+2}-1$; the indicated points end up at $(-3,-2)$ and $(-1, 0)$. The horizontal asymptote is now $y=-1$.

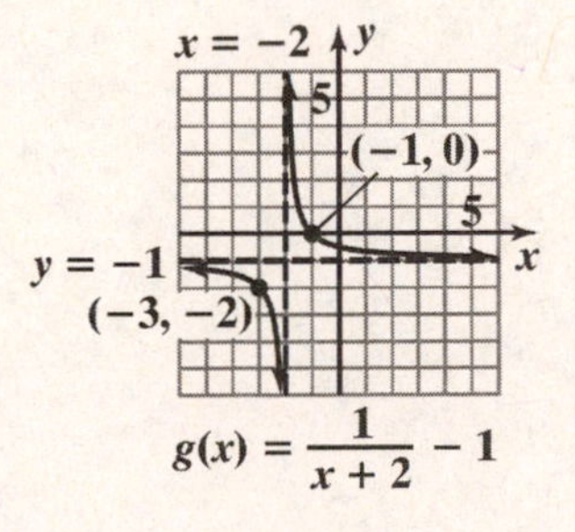

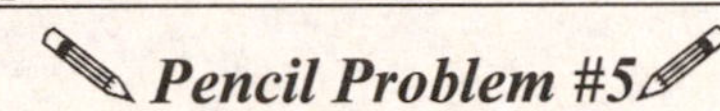

Pencil Problem #5

5. Use the graph of $f(x)=\dfrac{1}{x}$ to graph

$g(x)=\dfrac{1}{x+1}-2.$

Objective #6: Graph rational functions.

✔ Solved Problem #6

6a. Graph: $f(x)=\dfrac{3x-3}{x-2}$

Step 1: $f(-x)=\dfrac{3(-x)-3}{-x-2}=\dfrac{-3x-3}{-x-2}=\dfrac{3x+3}{x+2}$

Because $f(-x)$ does not equal $f(x)$ or $-f(x)$, the graph has neither y-axis symmetry nor origin symmetry.

Step 2: $f(0)=\dfrac{3(0)-3}{0-2}=\dfrac{3}{2}$

The y-intercept is $\dfrac{3}{2}$.

Step 3: $3x-3=0$

$3x=3$

$x=1$

The x-intercept is 1.

Step 4: $x-2=0$

$x=2$

The line $x=2$ is the only vertical asymptote for the graph of f.

Step 5: The numerator and denominator have the same degree, 1. The leading coefficients of the numerator and denominator are 3 and 1, respectively. Thus, the equation of the horizontal asymptote is $y=\dfrac{3}{1}$ or $y=3$.

Step 6: Plot points between and beyond each x-intercept and vertical asymptote:

x	-1	$\frac{3}{2}$	3	5
$f(x)$	2	-3	6	4

Step 7: Use the preceding information to graph the function.

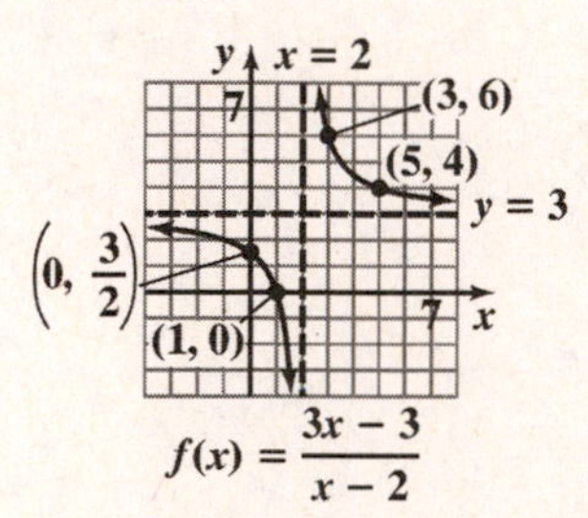

Pencil Problem #6

6a. Graph: $f(x)=\dfrac{-x}{x+1}$

6b. Graph: $f(x)=\dfrac{x^4}{x^2+2}$

Step 1: $f(-x)=\dfrac{(-x)^4}{(-x)^2+2}=\dfrac{x^4}{x^2+2}=f(x)$

Because $f(-x)=f(x)$, the graph has y-axis symmetry.

Step 2: $f(0)=\dfrac{0^4}{0^2+2}=0$

The y-intercept is 0, so the graph passes through the origin.

Step 3: $x^4=0$

$x=0$

There is only one x-intercept. This verifies that the graph passes through the origin.

Step 4: $x^2+2=0$

$x^2=-2$

$x=\pm i\sqrt{2}$

Since these solutions are not real, the graph of f will not have any vertical asymptotes.

Step 5: The degree of the numerator, 4, is greater than the degree of the denominator, 2, so the graph will not have a horizontal asymptote.

Step 6: Plot some points other than the intercepts:

x	-2	-1	1	2
$f(x)$	$\frac{8}{3}$	$\frac{1}{3}$	$\frac{8}{3}$	$\frac{1}{3}$

Step 7: Use the preceding information to graph the function.

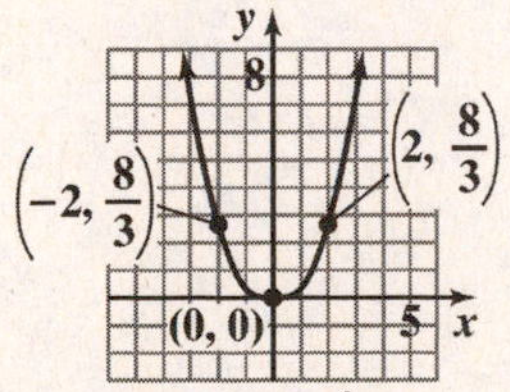

$\left(-2, \frac{8}{3}\right)$ $\left(2, \frac{8}{3}\right)$ $(0, 0)$

$f(x)=\dfrac{x^4}{x^2+2}$

6b. Graph: $f(x)=-\dfrac{1}{x^2-4}$

Objective #7: Identify slant asymptotes.

✔ *Solved Problem #7*

7. Find the slant asymptote of $f(x) = \dfrac{2x^2 - 5x + 7}{x - 2}$.

Note that the graph of *f* has a slant asymptote because the degree of the numerator is exactly one more than the degree of the denominator and the denominator is not a factor of the numerator.

Divide $2x^2 - 5x + 7$ by $x - 2$.

$$\begin{array}{r|rrr} 2 & 2 & -5 & 7 \\ & & 4 & -2 \\ \hline & 2 & -1 & 5 \end{array}$$

So, $\dfrac{2x^2 - 5x + 7}{x - 2} = 2x - 1 + \dfrac{5}{x - 2}$.

The equation of the slant asymptote is $y = 2x - 1$.

Pencil Problem #7

7. Find the slant asymptote of $f(x) = \dfrac{x^2 + x - 6}{x - 3}$.

Objective #8: Solve applied problems involving rational functions.

✔ *Solved Problem #8*

8. A company is planning to manufacture wheelchairs. The cost, *C*, in dollars, of producing *x* wheelchairs is $C(x) = 500{,}000 + 400x$.

8a. Write the average cost function, $\overline{C}$.

The average cost is the cost divided by the number of wheelchairs produced.

$$\overline{C}(x) = \frac{500{,}000 + 400x}{x}$$

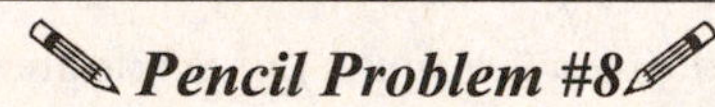

Pencil Problem #8

8. A company is planning to manufacture mountain bikes. The cost, *C*, in dollars, of producing *x* mountain bikes is $C(x) = 100{,}000 + 100x$.

8a. Write the average cost function, $\overline{C}$.

8b. Find and interpret $\bar{C}(1000)$ and $\bar{C}(10,000)$.

$$\bar{C}(1000) = \frac{500,000 + 400(1000)}{1000} = 900$$

The average cost per wheelchair of producing 1000 wheelchairs is $900.

$$\bar{C}(10,000) = \frac{500,000 + 400(10,000)}{10,000} = 405$$

The average cost per wheelchair of producing 10,000 wheelchairs is $405.

8c. What is the horizontal asymptote for the graph of $\bar{C}$? Describe what this means for the company.

The horizontal asymptote is $y = \frac{400}{1}$ or $y = 400$.

The cost per wheelchair approaches $400 as more wheelchairs are produced.

8b. Find and interpret $\bar{C}(1000)$ and $\bar{C}(4000)$.

8c. What is the horizontal asymptote for the graph of $\bar{C}$? Describe what this means for the company.

Answers for Pencil Problems *(Textbook Exercise references in parentheses)*:

1a. $\{x|x \neq -7, x \neq 7\}$ or $(-\infty, -7)\cup(-7, 7)\cup(7, \infty)$ *(3.5 #5)*

1b. all real numbers or $(-\infty, \infty)$ *(3.5 #7)*

2. True *(3.5 #14)*

3a. vertical asymptote: $x = -4$ *(3.5 #25)*

3b. no vertical asymptotes *(3.5 #27)*

4a. horizontal asymptote: $y = \frac{-2}{3}$ *(3.5 #43)*

4b. horizontal asymptote: $y = 0$ *(3.5 #37)*

5.

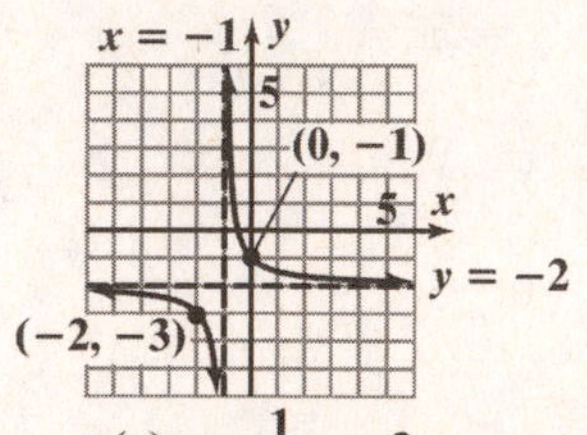

$g(x) = \frac{1}{x+1} - 2$ *(3.5 #49)*

6a.

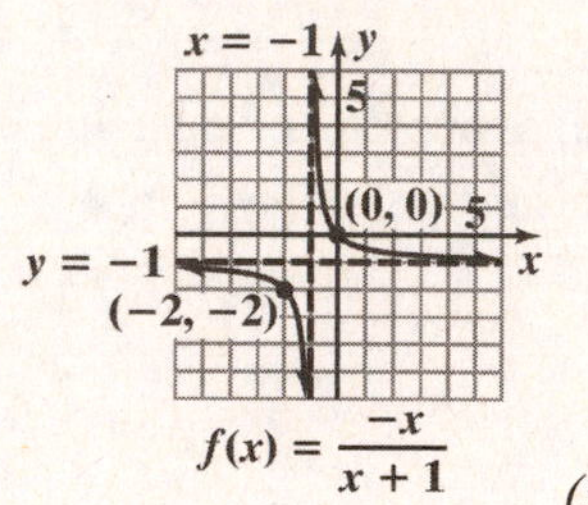

$f(x) = \frac{-x}{x+1}$ *(3.5 #63)*

6b.

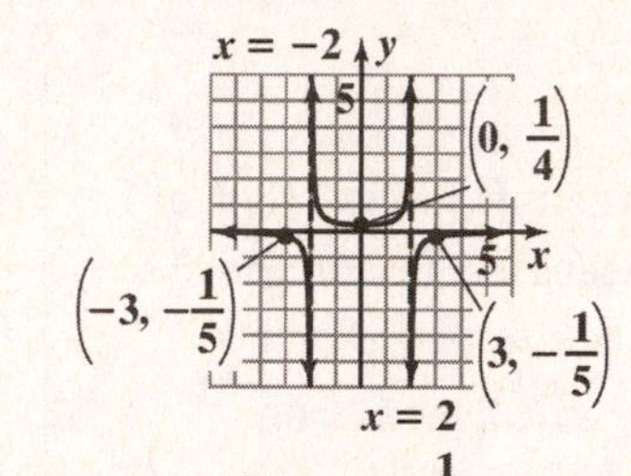

$f(x) = -\frac{1}{x^2 - 4}$ *(3.5 #65)*

7. $y = x + 4$ *(3.5 #85a)*

8a. $\overline{C}(x) = \frac{100{,}000 + 100x}{x}$ *(3.5 #99b)*

8b. $\overline{C}(1000) = 200$; The average cost per mountain bike of producing 1000 mountain bikes is $200; $\overline{C}(4000) = 125$; The average cost per mountain bike of producing 4000 mountain bikes is $125. *(3.5 #99c)*

8c. $y = 100$; The cost per mountain bike approaches $100 as more mountain bikes are produced. *(3.5 #99d)*

Section 3.6
Polynomial and Rational Inequalities

Tailgaters Beware!

It is never a good idea to follow too closely behind the car in front of you.
But when the roads are wet it can be even more dangerous.

In this section, we apply the mathematical concepts we learn to explore the different stopping distances required for a car driving on wet pavement and a car driving on dry pavement.

Objective #1: Solve polynomial inequalities.

✔ *Solved Problem #1*

1. Solve and graph the solution set on a real number line: $x^2 - x > 20$

$$x^2 - x > 20$$
$$x^2 - x - 20 > 0$$

Solve the related quadratic equation to find the boundary points.

$$x^2 - x - 20 = 0$$
$$(x+4)(x-5) = 0$$

Apply the zero-product principle.

$x + 4 = 0$ or $x - 5 = 0$

$x = -4$ $\quad\quad$ $x = 5$

The boundary points are -4 and 5.

Interval	Test Value	Test	Conclusion
$(-\infty,-4)$	-5	$(-5)^2 - (-5) > 20$ $30 > 20$, true	$(-\infty,-4)$ belongs to the solution set.
$(-4,5)$	0	$(0)^2 - (0) > 20$ $0 > 20$, false	$(-4,5)$ does not belong to the solution set.
$(5,\infty)$	10	$(10)^2 - (10) > 20$ $90 > 20$, true	$(5,\infty)$ belongs to the solution set.

The solution set is $(-\infty,-4) \cup (5,\infty)$.

−8 −6 −4 −2 0 2 4 6 8 10 12

Pencil Problem #1

1. Solve and graph the solution set on a real number line: $4x^2+7x<-3$

Objective #2: Solve rational inequalities.

✔ *Solved Problem #2*

2. Solve and graph the solution set on a real number line: $\frac{2x}{x+1} \ge 1$

$$\frac{2x}{x+1} \ge 1$$
$$\frac{2x}{x+1} - 1 \ge 0$$
$$\frac{2x}{x+1} - \frac{x+1}{x+1} \ge 0$$
$$\frac{2x - x - 1}{x+1} \ge 0$$
$$\frac{x-1}{x+1} \ge 0$$

Find the values of x that make the numerator and denominator zero.

$x - 1 = 0$ and $x + 1 = 0$

$x = 1$ $\qquad$ $x = -1$

The boundary points are -1 and 1.

Interval	Test Value	Test	Conclusion
$(-\infty, -1)$	-2	$\frac{2(-2)}{-2+1} \ge 1$ $4 \ge 1$, true	$(-\infty, -1)$ belongs to the solution set.
$(-1, 1)$	0	$\frac{2(0)}{0+1} \ge 1$ $0 \ge 1$, false	$(-1, 1)$ does not belong to the solution set.
$(1, \infty)$	2	$\frac{2(2)}{2+1} \ge 1$ $\frac{4}{3} \ge 1$, true	$(1, \infty)$ belongs to the solution set.

Exclude -1 from the solution set because it would make the denominator zero. The solution set is $(-\infty, -1) \cup [1, \infty)$.

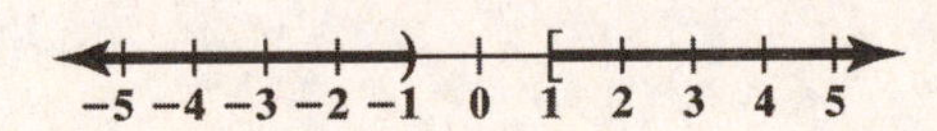

Pencil Problem #2

2. Solve and graph the solution set on a real number line: $\dfrac{x+1}{x+3} < 2$

Objective #3: Solve problems modeled by polynomial or rational inequalities.

✔ *Solved Problem #3*

3. An object is propelled straight up from ground level with an initial velocity of 80 feet per second. Its height at time t is modeled by $s(t) = -16t^2 + 80t$ where the height, $s(t)$, is measured in feet and the time, t, is measured in seconds. In which time interval will the object be more than 64 feet above the ground?

To find when the object will be more than 64 feet above the ground, solve the inequality $-16t^2 + 80t > 64$.
Solve the related quadratic equation.

$$-16t^2 + 80t = 64$$

$$-16t^2 + 80t - 64 = 0$$

$$t^2 - 5t + 4 = 0$$

$$(t-4)(t-1) = 0$$

$$t - 4 = 0 \quad \text{or} \quad t - 1 = 0$$

$$t = 4 \qquad\qquad t = 1$$

The boundary points are 1 and 4.

Interval	Test Value	Test	Conclusion
$(0,1)$	0.5	$-16(0.5)^2 + 80(0.5) > 64$ $36 > 64$, false	$(0,1)$ does not belong to the solution set.
$(1,4)$	2	$-16(2)^2 + 80(2) > 64$ $96 > 64$, true	$(1,4)$ belongs to the solution set.
$(4,\infty)$	5	$-16(5)^2 + 80(5) > 64$ $0 > 64$, false	$(4,\infty)$ does not belong to the solution set.

The solution set is $(1,4)$.

This means that the object will be more than 64 feet above the ground between 1 and 4 seconds excluding $t = 1$ and $t = 4$.

Pencil Problem #3

3. You throw a ball straight up from a rooftop 160 feet high with an initial speed of 48 feet per second. The function $s(t) = -16t^2 + 48t + 160$ models the ball's height above the ground, $s(t)$, in feet, t seconds after it was thrown. During which time period will the ball's height exceed that of the rooftop?

Answers for Pencil Problems *(Textbook Exercise references in parentheses)*:

1. $\left(-1,-\frac{3}{4}\right)$

−1 0 1 *(3.6 #15)*

2. $(-\infty,-5)\cup(-3,\infty)$

−6 −5 −4 −3 −2 −1 0 1 2 3 4 *(3.6 #55)*

3. The ball exceeds the height of the building between 0 and 3 seconds. *(3.6 #76)*

Section 3.7
Modeling Using Variation

How Far Would You Go To Lose Weight?

On the moon your weight would be significantly less.

To find out how much less,
be sure to work on the application problems
in this section of your textbook!

Objective #1: Solve direct variation problems.

✔ *Solved Problem #1*

1. The number of gallons of water, W, used when taking a shower varies directly as the time, t, in minutes, in the shower. A shower lasting 5 minutes uses 30 gallons of water. How much water is used in a shower lasting 11 minutes?

Since W varies directly with t, we have $W = kt$.
Use the given values to find k.

$$W = kt$$
$$30 = k \cdot 5$$
$$\frac{30}{5} = \frac{k \cdot 5}{5}$$
$$6 = k$$

The equation becomes $W = 6t$. Find W when $t = 11$.

$$W = 6t$$
$$W = 6 \cdot 11$$
$$= 66$$

An 11 minute shower will use 66 gallons of water.

Pencil Problem #1

1. An alligator's tail length, T, varies directly as its body length, B. An alligator with a body length of 4 feet has a tail length of 3.6 feet. What is the tail length of an alligator whose body length is 6 feet?

Objective #2: Solve inverse variation problems.

✔ Solved Problem #2

2. The length of a violin string varies inversely as the frequency of its vibrations. A violin string 8 inches long vibrates at a frequency of 640 cycles per second. What is the frequency of a 10-inch string?

Beginning with $y = \frac{k}{x}$, we will use l for the length of the string and f for the frequency.

Use the given values to find k.

$$f = \frac{k}{l}$$
$$640 = \frac{k}{8}$$
$$8 \cdot 640 = 8 \cdot \frac{k}{8}$$
$$5120 = k$$

The equation becomes $f = \frac{k}{l}$

$$f = \frac{5120}{l}$$

Find f when $l = 10$.

$$f = \frac{5120}{l}$$
$$f = \frac{5120}{10}$$
$$f = 512$$

A string length of 10 inches will vibrate at 512 cycles per second.

Pencil Problem #2

2. A bicyclist tips his bicycle when making a turn. The angle B, formed by the vertical direction and the bicycle, is called the banking angle. The banking angle varies inversely as the cycle's turning radius. When the turning radius is 4 feet, the banking angle is $28°$. What is the banking angle when the turning radius is 3.5 feet?

Objective #3: Solve combined variation problems.

✔ Solved Problem #3

3. The number of minutes needed to solve an Exercise Set of variation problems varies directly as the number of problems and inversely as the number of people working to solve the problems. It takes 4 people 32 minutes to solve 16 problems. How many minutes will it take 8 people to solve 24 problems?

Let $m =$ the number of minutes needed to solve an exercise set.
Let $p =$ the number of people working on the problems.
Let $x =$ the number of problems in the exercise set.

Use $m = \frac{kx}{p}$ to find k.

$$m = \frac{kx}{p}$$
$$32 = \frac{k16}{4}$$
$$32 = 4k$$
$$k = 8$$

Thus, $m = \frac{8x}{p}$.

Find m when $p = 8$ and $x = 24$.

$$m = \frac{8 \cdot 24}{8}$$
$$m = 24$$

It will take 24 minutes for 8 people to solve 24 problems.

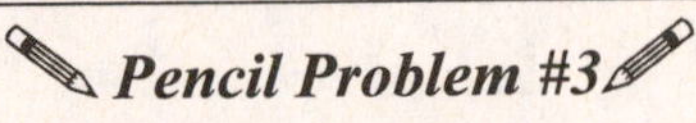

Pencil Problem #3

3. Body-mass index, or BMI, varies directly as one's weight, in pounds, and inversely as the square of one's height, in inches. A person who weighs 180 pounds and is 5 feet, or 60 inches, tall has a BMI of 35.15. What is the BMI, to the nearest tenth, for a 170 pound person who is 5 feet 10 inches tall?

Objective #4: Solve problems involving joint variation.

✔ *Solved Problem #4*

4. The volume of a cone, V, varies jointly as its height, h, and the square of its radius, r. A cone with a radius measuring 6 feet and a height measuring 10 feet has a volume of 120π cubic feet. Find the volume of a cone having a radius of 12 feet and a height of 2 feet.

Find k: $V = khr^2$

$$120\pi = k\cdot 10\cdot 6^2$$
$$120\pi = k\cdot 360$$
$$\frac{120\pi}{360} = \frac{k\cdot 360}{360}$$
$$\frac{\pi}{3} = k$$

Thus, $V = \frac{\pi}{3}hr^2 = \frac{\pi hr^2}{3}$.

$$V = \frac{\pi hr^2}{3}$$
$$V = \frac{\pi\cdot 2\cdot 12^2}{3} = 96\pi$$

The volume of a cone having a radius of 12 feet and a height of 2 feet is 96π cubic feet.

Pencil Problem #4

4. The heat loss of a glass window varies jointly as the window's area and the difference between the outside and inside temperatures. A window 3 feet wide by 6 feet long loses 1200 Btu per hour when the temperature outside is $20°$ colder than the temperature inside. Find the heat loss through a glass window that is 6 feet wide by 9 feet long when the temperature outside is $10°$ colder than the temperature inside.

Answers for Pencil Problems *(Textbook Exercise references in parentheses)*:

1. 5.4 feet *(3.7 #21)* **2.** 32° *(3.7 #27)* **3.** BMI: 24.4 *(3.7 #31)* **4.** 1800 Btu *(3.7 #33)*

Section 4.1
Exponential Functions

Shop 'til You Drop!

Are you just browsing? Take your time.
Researchers know, to the dollar, the average amount the typical consumer spends at the shopping mall. And the longer you stay, the more you spend. So if you say you're just browsing, that's just fine with the mall merchants. Browsing is time and, as we will explore in this section, time is money.

Objective #1: Evaluate exponential functions.

✔ *Solved Problem #1*

1. The exponential function $f(x) = 42.2(1.56)^x$ models the average amount spent, $f(x)$, in dollars, at a shopping mall after x hours. What is the average amount spent, to the nearest dollar, after three hours?

$f(x) = 42.2(1.56)^x$

$f(3) = 42.2(1.56)^3$

≈ 160.20876

≈ 160

The average amount spent after three hours at a mall is \$160.

Pencil Problem #1

1. The exponential function $f(x) = 574(1.026)^x$ models the population of India, $f(x)$, in millions, x years after 1974. Find India's population, to the nearest million, in the year 2028.

Objective #2: Graph exponential functions.

✔ *Solved Problem #2*

2a. Graph: $f(x) = 3^x$

Make a table of values:

x	$f(x) = 3^x$	(x, y)
-3	$3^{-3} = \frac{1}{27}$	$\left(-3, \frac{1}{27}\right)$
-2	$3^{-2} = \frac{1}{9}$	$\left(-2, \frac{1}{9}\right)$
-1	$3^{-1} = \frac{1}{3}$	$\left(-1, \frac{1}{3}\right)$
0	$3^0 = 1$	$(0, 1)$
1	$3^1 = 3$	$(1, 3)$
2	$3^2 = 9$	$(2, 9)$
3	$3^3 = 27$	$(3, 27)$

(Continued on next page)

Pencil Problem #2

2a. Graph: $f(x) = 4^x$

Plot the points in the table and connect with a smooth curve.

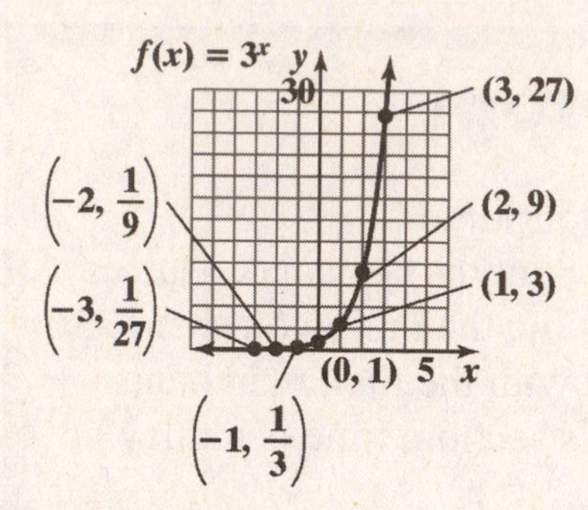

2b. Graph: $f(x)=\left(\frac{1}{3}\right)^x$

Make a table of values:

x	$f(x)=\left(\frac{1}{3}\right)^x$	(x, y)
-3	$\left(\frac{1}{3}\right)^{-3}=27$	$(-3, 27)$
-2	$\left(\frac{1}{3}\right)^{-2}=9$	$(-2, 9)$
-1	$\left(\frac{1}{3}\right)^{-1}=3$	$(-1, 3)$
0	$\left(\frac{1}{3}\right)^{0}=1$	$(0, 1)$
1	$\left(\frac{1}{3}\right)^{1}=\frac{1}{3}$	$\left(1,\frac{1}{3}\right)$
2	$\left(\frac{1}{3}\right)^{2}=\frac{1}{9}$	$\left(2,\frac{1}{9}\right)$
3	$\left(\frac{1}{3}\right)^{3}=\frac{1}{27}$	$\left(3,\frac{1}{27}\right)$

Plot the points in the table and connect with a smooth curve.

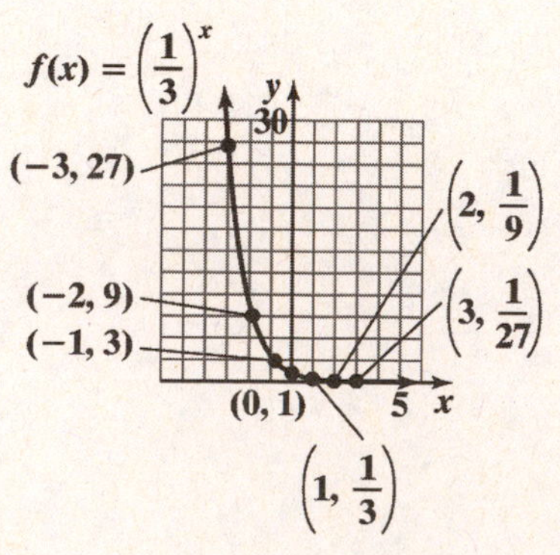

2b. Graph: $g(x)=\left(\frac{3}{2}\right)^x$

2c. Use the graph of $f(x) = 3^x$ to obtain the graph of $g(x) = 3^{x-1}$.

The graph of $g(x) = 3^{x-1}$ is the graph of $f(x) = 3^x$ shifted 1 unit to the right. We identified two points on the graph of *f*, which we graphed in Solved Problem 2a, and added 1 to each of the *x*-coordinates.

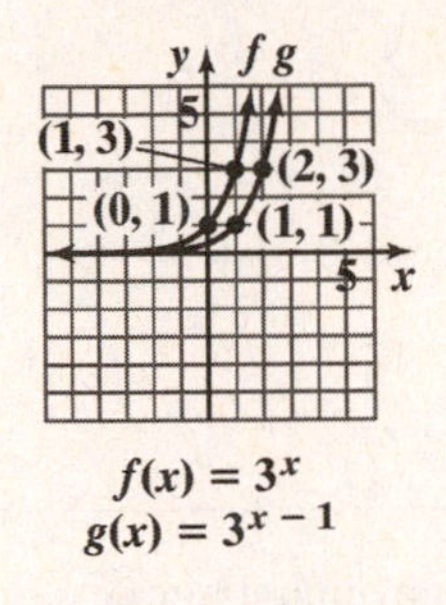

$f(x) = 3^x$
$g(x) = 3^{x-1}$

2c. Use the graph of $f(x) = 2^x$ to obtain the graph of $g(x) = 2^{x+1}$.

2d. Use the graph of $f(x) = 2^x$ to obtain the graph of $g(x) = 2^x + 1$.

The graph of $g(x) = 2^x + 1$ is the graph of $f(x) = 2^x$ shifted up 1 unit. We identified two points on the graph of *f* and added 1 to each of the *y*-coordinates. The asymptote is also shifted up 1 unit.

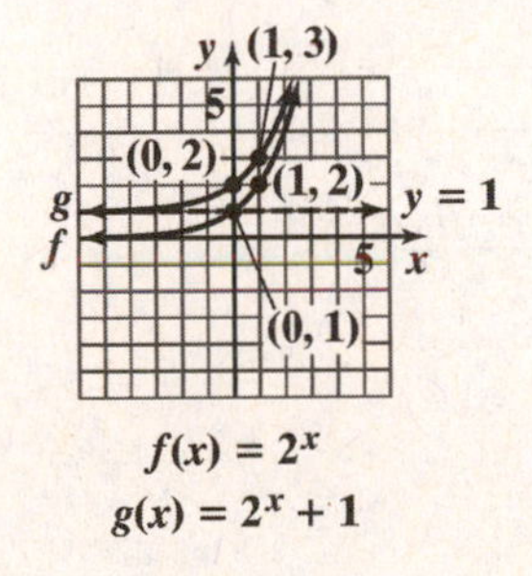

$f(x) = 2^x$
$g(x) = 2^x + 1$

2d. Use the graph of $f(x) = 2^x$ to obtain the graph of $g(x) = 2^x - 1$.

Objective #3: Evaluate functions with base *e*.

✔ *Solved Problem #3*

3. The exponential function $f(x) = 1066e^{0.042x}$ models the gray wolf population of the Western Great Lakes, $f(x)$, x years after 1978. If trends continue, project the gray wolf's population in the recovery area in 2012.

2012 is 34 years after 1978.

$$f(x) = 1066e^{0.042x}$$

$$f(34) = 1066e^{0.042(34)} \approx 4446$$

In 2012 the gray wolf population of the Western Great Lakes was projected to be about 4446.

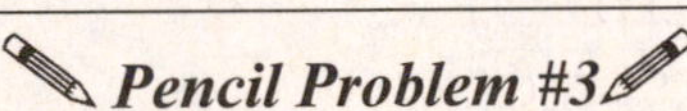

Pencil Problem #3

3. The exponential function $g(x) = 6875e^{0.077x}$ models the average cost of a family health insurance plan *x* years after 2000. What was the average cost of a family health insurance plan in 2011?

Objective #4: Use compound interest formulas.

✔ Solved Problem #4

4a. A sum of $10,000 is invested at an annual rate of 8%. Find the balance in the account after 5 years subject to quarterly compounding.

$$A = P\left(1+\frac{r}{n}\right)^{nt}$$

$$= \$10,000\left(1+\frac{0.08}{4}\right)^{4\cdot 5}$$

$$\approx \$14,859.47$$

4b. A sum of $10,000 is invested at an annual rate of 8%. Find the balance in the account after 5 years subject to continuous compounding.

$$A = Pe^{rt}$$

$$= \$10,000e^{0.08(5)}$$

$$\approx \$14,918.25$$

Pencil Problem #4

4a. A sum of $10,000 is invested at an annual rate of 5.5%. Find the balance in the account after 5 years subject to monthly compounding.

4b. A sum of $10,000 is invested at an annual rate of 5.5%. Find the balance in the account after 5 years subject to continuous compounding.

Answers for Pencil Problems *(Textbook Exercise references in parentheses)*:

1. 2295 million *(4.1 #65c)*

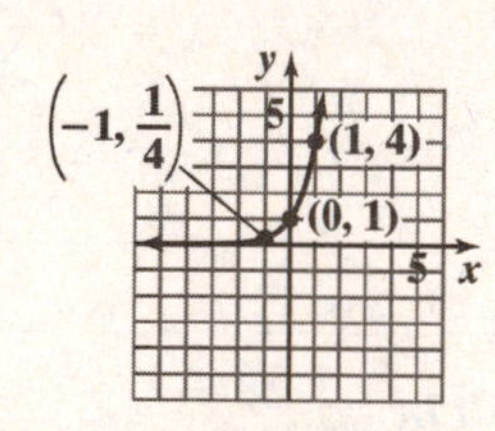

2a. $f(x) = 4^x$ *(4.1 #11)*

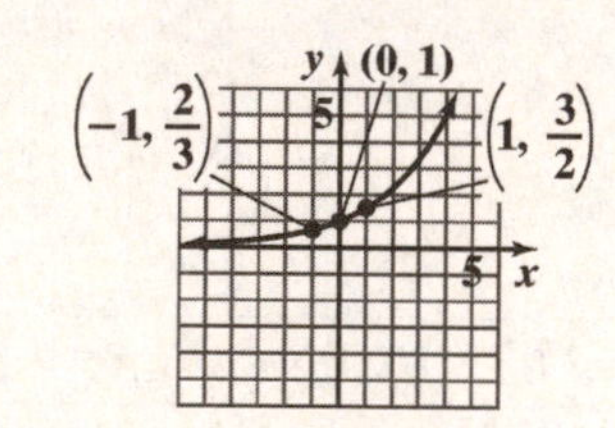

2b. $g(x) = \left(\frac{3}{2}\right)^x$ *(4.1 #13)*

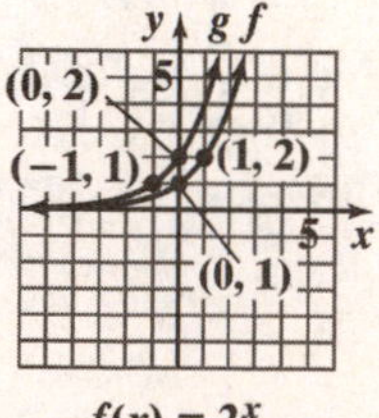

2c. $f(x) = 2^x$, $g(x) = 2^{x+1}$ *(4.1 #25)*

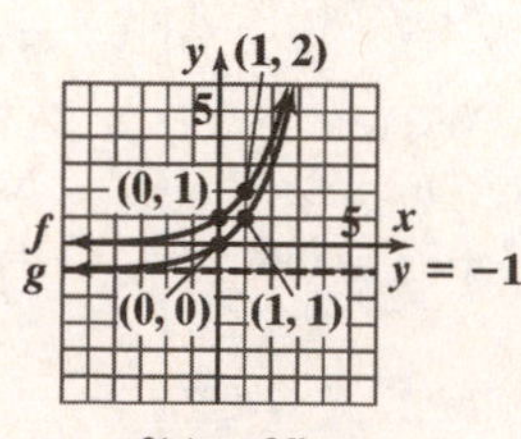

2d. $f(x) = 2^x$, $g(x) = 2^x - 1$ *(4.1 #27)*

3. $16,037 *(4.1 #71b)*

4a. $13,157.04 *(4.1 #53c)* **4b.** $13,165.31 *(4.1 #53d)*

Section 4.2
Logarithmic Functions

Speak Up!

The loudness level of a sound is measured in decibels. Decibel levels range from 0, a barely audible sound, to 160, a sound resulting in a ruptured eardrum.

We will see that decibels can be modeled by a logarithmic function, the topic of this section.

Objective #1: Change from logarithmic to exponential form.

✔ Solved Problem #1

1. Write the equation $2 = \log_b 25$ in its equivalent exponential form.

$2 = \log_b 25$

$b^2 = 25$

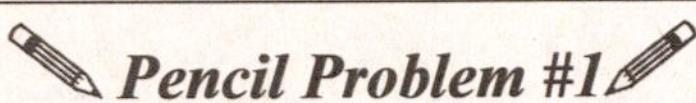

1. Write the equation $\log_6 216 = y$ in its equivalent exponential form.

Objective #2: Change from exponential to logarithmic form.

✔ Solved Problem #2

2. Write the equation $2^5 = x$ in its equivalent logarithmic form.

$2^5 = x$

$5 = \log_2 x$

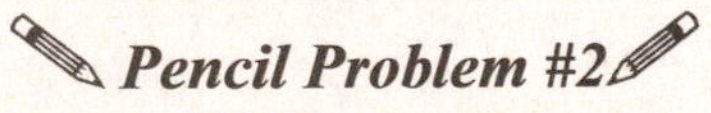

2. Write the equation $7^y = 200$ in its equivalent logarithmic form.

Objective #3: Evaluate logarithms.

✔ Solved Problem #3

3a. Evaluate: $\log_5 \frac{1}{125}$.

$\log_5 \frac{1}{125} = -3$ because

$5^{-3} = \frac{1}{5^3} = \frac{1}{125}$

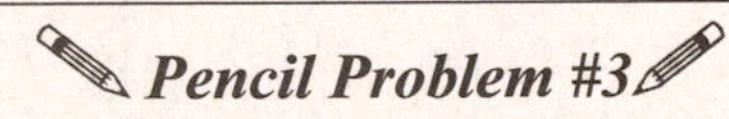

3. Evaluate: $\log_5 \frac{1}{5}$.

3b. Evaluate: $\log_{36} 6$.

$\log_{36} 6 = \frac{1}{2}$ because $36^{\frac{1}{2}} = \sqrt{36} = 6$.

3b. Evaluate: $\log_4 16$.

Objective #4: Use basic logarithmic properties.

✔ *Solved Problem #4*

4. Evaluate $\log_7 7^8$.

Because $\log_b b^x = x$, we conclude $\log_7 7^8 = 8$.

Pencil Problem #4

4. Evaluate $\log_4 1$.

Objective #5: Graph logarithmic functions.

✔ *Solved Problem #5*

5. Graph $f(x) = 3^x$ and $g(x) = \log_3 x$ in the same rectangular coordinate system.

Set up a table of coordinates for $f(x) = 3^x$.

x	-2	-1	0	1	2	3
$f(x) = 3^x$	$\frac{1}{9}$	$\frac{1}{3}$	1	3	9	27

Reverse these coordinates to obtain the coordinates of $g(x) = \log_3 x$.

x	$\frac{1}{9}$	$\frac{1}{3}$	1	3	9	27
$g(x) = \log_3 x$	-2	-1	0	1	2	3

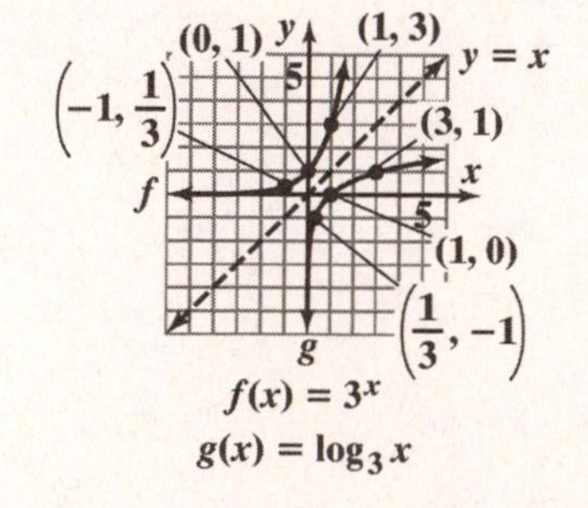

Pencil Problem #5

5. Graph $f(x) = \left(\frac{1}{2}\right)^x$ and $g(x) = \log_{\frac{1}{2}} x$ in the same rectangular coordinate system.

Objective #6: Find the domain of a logarithmic function.

✔ *Solved Problem #6*

6. Find the domain of $f(x) = \log_4(x-5)$.

$$x - 5 > 0$$
$$x > 5$$

The domain of *f* is $(5, \infty)$.

Pencil Problem #6

6. Find the domain of $f(x) = \log(2-x)$.

Objective #7: Use common logarithms.

✔ *Solved Problem #7*

7. The percentage of adult height attained by a boy who is *x* years old can be modeled by $f(x) = 29 + 48.8\log(x+1)$, where *x* represents the boy's age and $f(x)$ represents the percentage of his adult height. Approximately what percentage of his adult height has a boy attained at age ten?

$$f(x) = 29 + 48.8\log(x+1)$$
$$f(10) = 29 + 48.8\log(10+1)$$
$$= 29 + 48.8\log 11$$
$$\approx 80$$

A 10-year-old boy has attained approximately 80% of his adult height.

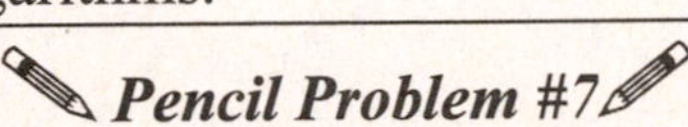

Pencil Problem #7

7. The percentage of adult height attained by a girl who is *x* years old can be modeled by $f(x) = 62 + 35\log(x-4)$, where *x* represents the girl's age and $f(x)$ represents the percentage of her adult height. Approximately what percentage of her adult height has a girl attained at age 13?

Objective #8: Use natural logarithms.

✔ *Solved Problem #8*

8a. Find the domain of $f(x) = \ln(4-x)$.

The domain of *f* consists of all *x* for which $4 - x > 0$.

$$4 - x > 0$$
$$-x > -4$$
$$x < 4$$

The domain of *f* is $(-\infty, 4)$.

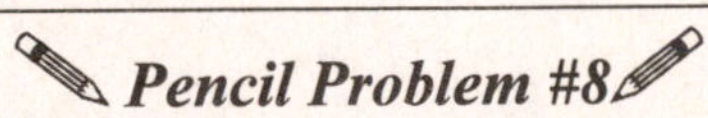

Pencil Problem #8

8a. Find the domain of $f(x) = \ln(x+2)$.

8b. The function $f(x) = 13.4\ln x - 11.6$ models the temperature increase, $f(x)$, in an enclosed vehicle after x minutes when the outside air temperature is between 72°F and 96°F. Use the function to find the temperature increase, to the nearest degree, after 30 minutes.

Evaluate the function at 30.

$$f(x) = 13.4\ln x - 11.6$$
$$f(30) = 13.4\ln 30 - 11.6$$
$$\approx 34$$

The temperature will increase by about 34° after 30 minutes.

8b. The function $f(x) = -7.2\ln x + 53$ models the percentage of first-year college men expressing antifeminist views x years after 1969. Use the function to project the percentage of first-year college men who will express antifeminist views in 2015. Round to one decimal place.

Answers for Pencil Problems *(Textbook Exercise references in parentheses)*:

1. $6^y = 216$ *(4.2 #7)*

2. $\log_7 200 = y$ *(4.2 #19)*

3a. -1 *(4.2 #21)* **3b.** 2 *(4.2 #21)*

4. 0 *(4.2 #37)*

5. 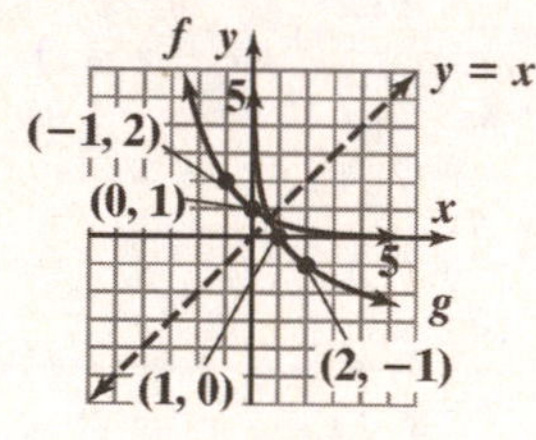

$f(x) = \left(\frac{1}{2}\right)^x$

$g(x) = \log_{1/2} x$

(4.2 #45)

6. $(-\infty, 2)$ *(4.2 #49)*

7. 95.4% *(4.2 #87)*

8a. $(-2, \infty)$ *(4.2 #81)*

8b. 24.2% *(4.2 #115)*

Section 4.3
Properties of Logarithms

How Smart Is This Chimp?

Scientists are often amazed at what a chimpanzee can learn. These scientists are not simply interested in *what* a chimp can learn, but also in a particular chimp's *maximum capacity* for learning, and *how long* the learning takes.

A typical chimpanzee learning sign language can master a maximum of 65 signs. In the Exercise Set of the textbook, one application will take a look at the number of weeks it will take for a chimp to master 30 signs.

Objective #1: Use the product rule.

✔ ***Solved Problem #1***

1. Use the product rule to expand: $\log(100x)$.

$$\log(100x) = \log 100 + \log x$$
$$= \log 10^2 + \log x$$
$$= 2 + \log x$$

Pencil Problem #1

1. Use the product rule to expand: $\log_5(7 \cdot 3)$.

Objective #2: Use the quotient rule.

✔ ***Solved Problem #2***

2. Use the quotient rule to expand: $\ln\left(\frac{e^5}{11}\right)$.

$$\ln\left(\frac{e^5}{11}\right) = \ln e^5 - \ln 11$$
$$= 5 - \ln 11$$

Pencil Problem #2

2. Use the quotient rule to expand: $\log_4\left(\frac{64}{y}\right)$.

Objective #3: Use the power rule.

✔ ***Solved Problem #3***

3a. Use the power rule to expand: $\log_6 3^9$

$$\log_6 3^9 = 9\log_6 3$$

Pencil Problem #3

3a. Use the power rule to expand: $\log_b x^3$.

3b. Use the power rule to expand: $\log(x+4)^2$

$\log(x+4)^2 = 2\log(x+4)$

3b. Use the power rule to expand: $\log N^{-6}$

Objective #4: Expand logarithmic expressions.

✔ *Solved Problem #4*

4a. Use logarithmic properties to expand: $\ln\sqrt[3]{x}$.

$$\ln\sqrt[3]{x} = \ln x^{\frac{1}{3}}$$
$$= \tfrac{1}{3}\ln x$$

4b. Use logarithmic properties to expand: $\log_b\left(x^4\sqrt[3]{y}\right)$.

$$\log_b\left(x^4\sqrt[3]{y}\right) = \log_b x^4 + \log_b \sqrt[3]{y}$$
$$= \log_b x^4 + \log_b y^{\frac{1}{3}}$$
$$= 4\log_b x + \tfrac{1}{3}\log_b y$$

4c. Use logarithmic properties to expand: $\log_5\left(\frac{\sqrt{x}}{25y^3}\right)$.

$$\log_5\left(\frac{\sqrt{x}}{25y^3}\right) = \log_5\left(\frac{x^{\frac{1}{2}}}{25y^3}\right)$$
$$= \log_5 x^{\frac{1}{2}} - \log_5\left(25y^3\right)$$
$$= \log_5 x^{\frac{1}{2}} - \left(\log_5 25 + \log_5 y^3\right)$$
$$= \log_5 x^{\frac{1}{2}} - \log_5 25 - \log_5 y^3$$
$$= \frac{1}{2}\log_5 x - 2 - 3\log_5 y$$

Pencil Problem #4

4a. Use logarithmic properties to expand: $\ln\sqrt[5]{x}$.

4b. Use logarithmic properties to expand: $\log_b(x^2 y)$.

4c. Use logarithmic properties to expand: $\log_6\left(\frac{36}{\sqrt{x+1}}\right)$.

Objective #5: Condense logarithmic expressions.

✔ *Solved Problem #5*

5a. Write as a single logarithm: $\log 25 + \log 4$.

$$\begin{aligned}\log 25 + \log 4 &= \log(25 \cdot 4)\\ &= \log 100\\ &= 2\end{aligned}$$

5b. Write as a single logarithm: $2\ln x + \frac{1}{3}\ln(x+5)$.

$$\begin{aligned}2\ln x + \frac{1}{3}\ln(x+5) &= \ln x^2 + \ln(x+5)^{\frac{1}{3}}\\ &= \ln x^2 + \ln \sqrt[3]{x+5}\\ &= \ln\left(x^2 \sqrt[3]{x+5}\right)\end{aligned}$$

5c. Write as a single logarithm:
$\frac{1}{4}\log_b x - 2\log_b 5 - 10\log_b y$.

$$\begin{aligned}&\tfrac{1}{4}\log_b x - 2\log_b 5 - 10\log_b y\\ &= \log_b x^{\frac{1}{4}} - \log_b 5^2 - \log_b y^{10}\\ &= \log_b x^{\frac{1}{4}} - \left(\log_b 5^2 + \log_b y^{10}\right)\\ &= \log_b \sqrt[4]{x} - \left(\log_b 25 + \log_b y^{10}\right)\\ &= \log_b \sqrt[4]{x} - \left(\log_b 25y^{10}\right)\\ &= \log_b\left(\frac{\sqrt[4]{x}}{25y^{10}}\right)\end{aligned}$$

✎ *Pencil Problem #5* ✎

5a. Write as a single logarithm: $\log 5 + \log 2$.

5b. Write as a single logarithm: $4\ln(x+6) - 3\ln x$.

5c. Write as a single logarithm:
$3\ln x + 5\ln y - 6\ln z$.

Objective #6: Use the change-of-base property.

Solved Problem #6

6a. Use common logarithms to evaluate $\log_7 2506$.

$$\log_7 2506 = \frac{\log 2506}{\log 7} \approx 4.02$$

6b. Use natural logarithms to evaluate $\log_7 2506$.

$$\log_7 2506 = \frac{\ln 2506}{\ln 7} \approx 4.02$$

Pencil Problem #6

6a. Use common logarithms to evaluate $\log_{0.1} 17$.

6b. Use natural logarithms to evaluate $\log_{0.1} 17$.

Answers for Pencil Problems *(Textbook Exercise references in parentheses)*:

1. $\log_5 7 + \log_5 3$ *(4.3 #1)*

2. $3 - \log_4 y$ *(4.3 #11)*

3a. $3\log_b x$ *(4.3 #15)* **3b.** $-6\log N$ *(4.3 #17)*

4a. $\frac{1}{5}\ln x$ *(4.3 #19)* **4b.** $2\log_b x + \log_b y$ *(4.3 #21)* **4c.** $2 - \frac{1}{2}\log_6(x+1)$ *(4.3 #25)*

5a. 1 *(4.3 #41)* **5b.** $\ln\left[\frac{(x+6)^4}{x^3}\right]$ *(4.3 #59)* **5c.** $\ln\left(\frac{x^3 y^5}{z^6}\right)$ *(4.3 #61)*

6a. -1.2304 *(4.3 #75)* **6b.** -1.2304 *(4.3 #75)*

Section 4.4
Exponential and Logarithmic Equations

Under the Sea!

Though it can be pitch black in the depths of the ocean, sunlight is visible as you get closer to the surface. About 12% of the surface sunlight reaches a depth of 20 feet and about 1.6% reaches to a depth of 100 feet.

In the applications of this section, you will use an exponential function to determine the depths that correspond to various percentages of light.

Objective #1: Use like bases to solve exponential equations.

✔ *Solved Problem #1*

1a. Solve: $5^{3x-6} = 125$.

$$5^{3x-6} = 125$$
$$5^{3x-6} = 5^3$$
$$3x - 6 = 3$$
$$3x = 9$$
$$x = 3$$

The solution set is $\{3\}$.

1b. Solve: $8^{x+2} = 4^{x-3}$.

$$8^{x+2} = 4^{x-3}$$
$$(2^3)^{x+2} = (2^2)^{x-3}$$
$$2^{3(x+2)} = 2^{2(x-3)}$$
$$3(x+2) = 2(x-3)$$
$$3x + 6 = 2x - 6$$
$$x + 6 = -6$$
$$x = -12$$

The solution set is $\{-12\}$.

Pencil Problem #1

1a. Solve: $4^{2x-1} = 64$.

1b. Solve: $8^{x+3} = 16^{x-1}$.

Objective #2: Use logarithms to solve exponential equations.

Solved Problem #2

2a. Solve: $10^x = 8000.$

Take the common log of both sides of the equation.

$$10^x = 8000$$
$$\log 10^x = \log 8000$$
$$x = \log 8000$$
$$x \approx 3.90$$

The solution set is $\{\log 8000 \approx 3.90\}$.

2b. Solve: $7e^{2x} - 5 = 58.$

Isolate the exponential expression, then take the natural log of both sides of the equation.

$$7e^{2x} - 5 = 58$$
$$7e^{2x} = 63$$
$$e^{2x} = 9$$
$$\ln e^{2x} = \ln 9$$
$$2x = \ln 9$$
$$x = \frac{\ln 9}{2}$$
$$x = \frac{\ln 3^2}{2}$$
$$x = \frac{2\ln 3}{2}$$
$$x = \ln 3$$
$$x \approx 1.10$$

The solution set is $\{\ln 3 \approx 1.10\}$.

Pencil Problem #2

2a. Solve: $10^x = 3.91.$

2b. Solve: $7^{x+2} = 410.$

Objective #3: Use exponential form to solve logarithmic equations.

Solved Problem #3

3a. Solve: $4\ln(3x)=8$.

$$4\ln(3x)=8$$
$$\ln(3x)=2$$
$$e^2=3x$$
$$\frac{e^2}{3}=x$$

Check:

$$4\ln(3x)=8$$
$$4\ln\left[3\left(\frac{e^2}{3}\right)\right]=8$$
$$4\ln\left(e^2\right)=8$$
$$4\cdot 2=8$$
$$8=8,\ \text{true}$$

The solution set is $\left\{\frac{e^2}{3}\right\}$.

3b. Solve: $\log x+\log(x-3)=1$.

$$\log x+\log(x-3)=1$$
$$\log(x^2-3x)=1$$
$$10^1=x^2-3x$$
$$0=x^2-3x-10$$
$$0=(x+2)(x-5)$$

$x+2=0$ or $x-5=0$

$x=-2$ $\quad x=5$

Check -2:

$$\log x+\log(x-3)=1$$
$$\log(-2)+\log(-2-3)=1$$

-2 does not check.

Check 5:

$$\log x+\log(x-3)=1$$
$$\log 5+\log(5-3)=1$$
$$\log 5+\log 2=1$$
$$\log 10=1$$
$$1=1,\ \text{true}$$

The solution set is $\{5\}$.

Pencil Problem #3

3a. Solve: $\log_4(x+5)=3$.

3b. Solve: $\log_5 x+\log_5(4x-1)=1$.

Objective #4: Use the one-to-one property of logarithms to solve logarithmic equations.

✔ Solved Problem #4

4. Solve: $\ln(x-3)=\ln(7x-23)-\ln(x+1)$

$$\ln(x-3)=\ln(7x-23)-\ln(x+1)$$
$$\ln(x-3)=\ln\left(\frac{7x-23}{x+1}\right)$$
$$x-3=\frac{7x-23}{x+1}$$
$$(x+1)(x-3)=(x+1)\frac{7x-23}{x+1}$$
$$x^2-2x-3=7x-23$$
$$x^2-9x+20=0$$
$$(x-4)(x-5)=0$$
$$x-4=0 \quad \text{or} \quad x-5=0$$
$$x=4 \qquad\qquad x=5$$

Check 4:
$$\ln(x-3)=\ln(7x-23)-\ln(x+1)$$
$$\ln(4-3)=\ln(7\cdot4-23)-\ln(4+1)$$
$$\ln 1=\ln 5-\ln 5$$
$$0=0,\ \text{true}$$

Check 5:
$$\ln(x-3)=\ln(7x-23)-\ln(x+1)$$
$$\ln(5-3)=\ln(7\cdot5-23)-\ln(5+1)$$
$$\ln 2=\ln 12-\ln 6$$
$$\ln 2=\ln\left(\frac{12}{6}\right)$$
$$\ln 2=\ln 2,\ \text{true}$$

The solution set is $\{4,5\}$.

Pencil Problem #4

4. Solve: $\log(x+4)-\log 2=\log(5x+1)$

Objective #5: Solve applied problems involving exponential and logarithmic equations.

✔ *Solved Problem #5*

5a. Medical research indicates that the risk of having a car accident increases exponentially as the concentration of alcohol in the blood increases. The risk is modeled by $R = 6e^{12.77x}$ where x is the blood alcohol concentration and R, given as a percent, is the risk of having a car accident. What blood alcohol concentration corresponds to a 7% risk of a car accident?

$$R = 6e^{12.77x}$$
$$6e^{12.77x} = 7$$
$$e^{12.77x} = \frac{7}{6}$$
$$\ln e^{12.77x} = \ln\frac{7}{6}$$
$$12.77x = \ln\frac{7}{6}$$
$$x = \frac{\ln\frac{7}{6}}{12.77}$$
$$x \approx 0.01$$

For a blood alcohol concentration of 0.01, the risk of a car accident is 7%.

Pencil Problem #5

5a. The formula $A = 37.3e^{0.0095e}$ models the population of California, A, in millions, t years after 2010. When will the population of California reach 40 million?

Solved Problem #5

5b. How long, to the nearest tenth of a year, will it take \$1000 to grow to \$3600 at 8% annual interest compounded quarterly?

$$A = P\left(1+\frac{r}{n}\right)^{nt}$$
$$3600 = 1000\left(1+\frac{0.08}{4}\right)^{4t}$$
$$3.6 = 1.02^{4t}$$
$$1.02^{4t} = 3.6$$
$$\ln 1.02^{4t} = \ln 3.6$$
$$4t\ln 1.02 = \ln 3.6$$
$$t = \frac{\ln 3.6}{4\ln 1.02}$$
$$t \approx 16.2$$

After approximately 16.2 years, the \$1000 will grow to \$3600.

Pencil Problem #5

5b. How long, to the nearest tenth of a year, will it take \$8000 to grow to \$16,000 at 8% annual interest compounded continuously?

Answers for Pencil Problems *(Textbook Exercise references in parentheses)*:

1a. $\{a\}$ *(4.4 #7)* **1b.** $\{13\}$ *(4.4 #19)*

2a. $\{\log 3.91 \approx 0.59\}$ *(4.4 #23)*

2b. $\left\{\dfrac{\ln 410}{\ln 7} - 2 \approx 1.09\right\}$ *(4.4 #37)*

3a. $\{59\}$ *(4.4 #53)*

3b. $\left\{\dfrac{5}{4}\right\}$; note: reject -1 *(4.4 #67)*

4. $\left\{\dfrac{2}{9}\right\}$ *(4.4 #83)*

5a. 2017 *(4.4 #103b)*

5b. 8.7 years *(4.4 #111)*

Section 4.5
Exponential Growth and Decay; Modeling Data

NOT Tooth Decay!

One of algebra's many applications is to predict the behavior of variables. This can be done with *exponential growth* and *decay models*. With exponential growth or decay, quantities grow or decay at a rate directly proportional to their size. Populations that are growing exponentially grow extremely rapidly as they get larger because there are more adults to have offspring.

In this section we explore how to create such functions and how to use them to make predictions.

Objective #1: Model exponential growth and decay.

✔ *Solved Problem #1*

1. In 2000, the population of Africa was 807 million and by 2011 it had grown to 1052 million.

1a. Use the exponential growth model $A = A_0e^{kt}$, in which t is the number of years after 2000, to find the exponential growth function that models the data.

2011 is 11 years after 2000.
Thus, when $t = 11$, $A = 1052$.
$A_0 = 807$

Substitute these values to find k.

$$A = A_0e^{kt}$$
$$1052 = 807e^{k(11)}$$

Solve for k.

$$1052 = 807e^{k(11)}$$
$$\frac{1052}{807} = e^{11k}$$
$$\ln \frac{1052}{807} = \ln e^{11k}$$
$$k = \frac{\ln \frac{1052}{807}}{11}$$
$$k \approx 0.024$$

Thus, the growth function is $A = 807e^{0.024t}$.

Pencil Problem #1

1. In 2000, the population of Israel was approximately 6.04 million and by 2050 it is projected to grow to 10 million.

1a. Use the exponential growth model $A = A_0e^{kt}$, in which t is the number of years after 2000, to find an exponential growth function that models the data.

1b. By which year will Africa's population reach 2000 million, or two billion?

$$A = 807e^{0.024t}$$
$$2000 = 807e^{0.024t}$$
$$\frac{2000}{807} = e^{0.024t}$$
$$\ln\frac{2000}{807} = \ln e^{0.024t}$$
$$\ln\frac{2000}{807} = 0.024t$$
$$t = \frac{\ln\frac{2000}{807}}{0.024} \approx 38$$

Africa's population will reach 2000 million approximately 38 years after 2000, or 2038.

1b. In which year will Israel's population be 9 million?

Objective #2: Use logistic growth models.

✔ *Solved Problem #2*

2. In a learning theory project, psychologists discovered that

$$f(t) = \frac{0.8}{1+e^{-0.2t}}$$

is a model for describing the proportion of correct responses, $f(t)$, after t learning trials.

2a. Find the proportion of correct responses after 10 learning trials.

We substitute 10 for t in the logistic growth function.

$$f(10) = \frac{0.8}{1+e^{-0.2(10)}} \approx 0.7$$

The proportion of correct responses after 10 trials is approximately 0.7.

✎ *Pencil Problem #2* ✎

2. The logistic growth function

$$f(t) = \frac{100{,}000}{1+5000e^{-t}}$$

describes the number of people, $f(t)$, who have become ill with influenza t weeks after its initial outbreak in a particular community.

2a. How many people were ill by the end of the fourth week?

2b. What is the limiting size of $f(t)$, the proportion of correct responses, as continued learning trials take place?

The number in the numerator of the logistic growth function, 0.8, is the limiting size of the proportion of correct responses.

2b. What is the limiting size of the population that becomes ill?

Objective #3: Choose an appropriate model for data.

✔ *Solved Problem #3*

3. The table shows the populations of various cities and the average walking speed of a person living in the city. Create a scatter plot for the data. Based on the scatter plot, what type of function would be a good choice for modeling the data?

Population (thousands)	Walking Speed (feet per second)
5.5	0.6
14	1.0
71	1.6
138	1.9
342	2.2

Scatter plot:

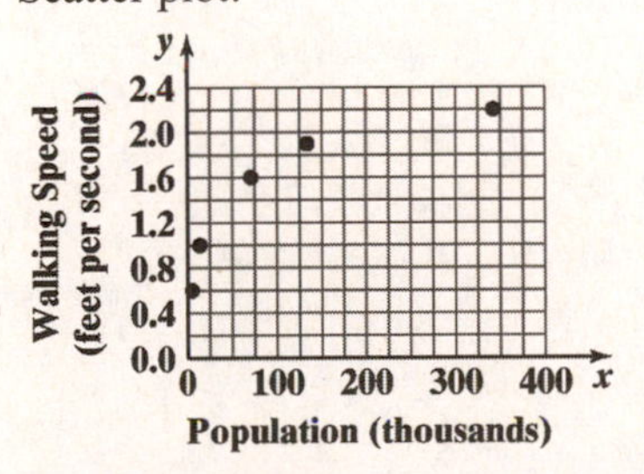

A logarithmic function would be a good choice for modeling the data.

Pencil Problem #3

3. The table shows the percentage of miscarriages by women of various ages. Create a scatter plot for the data. Determine the type of function that would be a good choice for modeling the data?

Woman's Age	Percent of Miscarriages
22	9%
27	10%
32	13%
37	20%
42	38%
47	52%

Objective #4: Express an exponential model in base *e*.

✔ Solved Problem #4

4. Rewrite $y = 4(7.8)^x$ in terms of base e. Express the answer in terms of a natural logarithm and then round to three decimal places.

$y = 4(7.8)^x$
$= 4e^{(\ln 7.8)x}$

Rounded to three decimal places:
$y = 4e^{2.054x}$

Pencil Problem #4

4. Rewrite $y = 100(4.6)^x$ in terms of base e. Express the answer in terms of a natural logarithm and then round to three decimal places.

Answers for Pencil Problems *(Textbook Exercise references in parentheses)*:

1a. $A = 6.04e^{0.01t}$ *(4.5 #7a)*

1b. 2040 *(4.5 #7b)*

2a. approximately 1080 people *(4.5 #37b)*

2b. 100,000 people *(4.5 #37c)*

3. 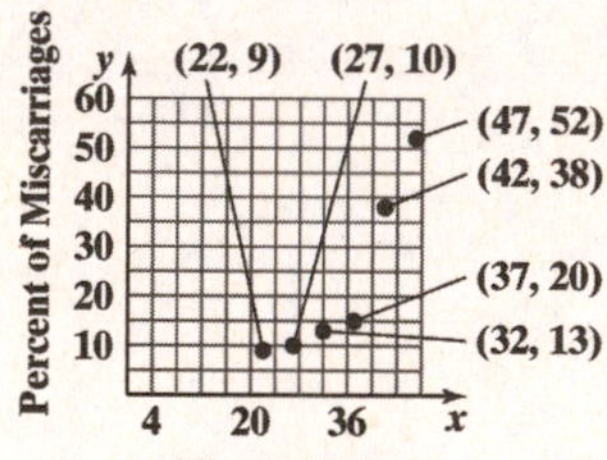

exponential function *(12.5 #47)*

4. $y = 100e^{(\ln 4.6)x} = 100e^{1.526x}$ *(4.5 #53)*

Section 5.1
Systems of Linear Equations in Two Variables

Procrastination makes you sick!

Researchers compared college students who were procrastinators and nonprocrastinators. Early in the semester, procrastinators reported fewer symptoms of illness, but late in the semester, they reported more symptoms than their nonprocrastinating peers.

In this section of the textbook, you will identify when both groups have the same number of symptoms as the point of intersection of two lines.

Objective #1: Decide whether an ordered pair is a solution of a linear system.

✔ *Solved Problem #1*

1. Determine if the ordered pair $(7,6)$ is a solution of the system: $\begin{cases} 2x-3y=-4 \\ 2x+y=4 \end{cases}$

To determine if $(7,6)$ is a solution to the system, replace x with 7 and y with 6 in both equations.

$$\begin{aligned} 2x-3y&=-4 \\ 2(7)-3(6)&=-4 \\ 14-18&=-4 \\ -4&=-4, \text{ true} \end{aligned} \qquad \begin{aligned} 2x+y&=4 \\ 2(7)+6&=4 \\ 14+6&=4 \\ 20&=4, \text{ false} \end{aligned}$$

The ordered pair does not satisfy both equations, so it is not a solution to the system.

Pencil Problem #1

1. Determine if the ordered pair $(2,3)$ is a solution of the system: $\begin{cases} x+3y=11 \\ x-5y=-13 \end{cases}$

Objective #2: Solve linear systems by substitution.

✔ *Solved Problem #2*

2. Solve by the substitution method: $\begin{cases} 3x+2y=4 \\ 2x+y=1 \end{cases}$

Solve $2x+y=1$ for y.

$$\begin{aligned} 2x+y&=1 \\ y&=-2x+1 \end{aligned}$$

(Continued on next page)

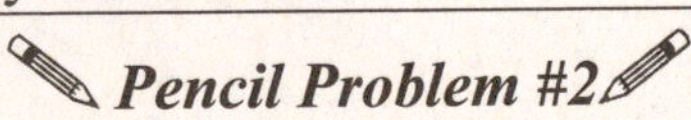

Pencil Problem #2

2. Solve by the substitution method: $\begin{cases} x+y=4 \\ y=3x \end{cases}$

Substitute: $3x + 2y = 4$

$$3x + 2(\overbrace{-2x+1}^{y}) = 4$$
$$3x - 4x + 2 = 4$$
$$-x + 2 = 4$$
$$-x = 2$$
$$x = -2$$

Find y.

$y = -2x + 1$

$y = -2(-2) + 1$

$y = 5$

The solution is $(-2, 5)$.

The solution set is $\{(-2, 5)\}$.

Objective #3: Solve linear systems by addition.

✔ *Solved Problem #3*

3. Solve the system: $\begin{cases} 4x + 5y = 3 \\ 2x - 3y = 7 \end{cases}$

Multiply each term of the second equation by –2 and add the equations to eliminate x.

$$\begin{array}{r} 4x + 5y = 3 \\ -4x + 6y = -14 \\ \hline 11y = -11 \\ y = -1 \end{array}$$

Back-substitute into either of the original equations to solve for x.

$$2x - 3y = 7$$
$$2x - 3(-1) = 7$$
$$2x + 3 = 7$$
$$2x = 4$$
$$x = 2$$

The solution set is $\{(2, -1)\}$.

Pencil Problem #3

3. Solve the system: $\begin{cases} 3x - 4y = 11 \\ 2x + 3y = -4 \end{cases}$

Objective #4: Identify systems that do not have exactly one ordered-pair solution.

✔ *Solved Problem #4*

4a. Solve the system: $\begin{cases} 5x - 2y = 4 \\ -10x + 4y = 7 \end{cases}$

Multiply the first equation by 2, and then add the equations.

$$\begin{array}{r} 10x - 4y = 8 \\ \underline{-10x + 4y = 7} \\ 0 = 15 \end{array}$$

Since there are no pairs (x, y) for which 0 will equal 15, the system is inconsistent and has no solution.
The solution set is $\varnothing$ or $\{\ \}$.

4b. Solve the system: $\begin{cases} x = 4y - 8 \\ 5x - 20y = -40 \end{cases}$

Substitute $4y - 8$ for x in the second equation.

$$\begin{aligned} 5x - 20y &= -40 \\ 5(\overbrace{4y-8}^{x}) - 20y &= -40 \\ 20y - 40 - 20y &= -40 \\ -40 &= -40 \end{aligned}$$

Since $-40 = -40$ for all values of x and y, the system is dependent.

The solution set is $\{(x, y) | x = 4y - 8\}$ or $\{(x, y) | 5x - 20y = -40\}$.

Pencil Problem #4

4a. Solve the system: $\begin{cases} x = 9 - 2y \\ x + 2y = 13 \end{cases}$

4b. Solve the system: $\begin{cases} y = 3x - 5 \\ 21x - 35 = 7y \end{cases}$

Objective #5: Solve problems using systems of linear equations.

✔ *Solved Problem #5*

5. A company that manufactures running shoes has a fixed cost of \$300,000. Additionally, it costs \$30 to produce each pair of shoes. The shoes are sold at \$80 per pair.

5a. Write the cost function, C, of producing x pairs of running shoes.

$$C(x) = \overbrace{300{,}000}^{\text{fixed costs}} + \overbrace{30x}^{\text{\$30 per pair}}$$

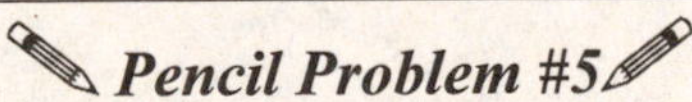

Pencil Problem #5

5. A company that manufactures small canoes has a fixed cost of \$18,000. Additionally, it costs \$20 to produce each canoe. The selling price is \$80 per canoe.

5a. Write the cost function, C, of producing x canoes.

5b. Write the revenue function, R, from the sale of x pairs of running shoes.

$$R(x) = \overbrace{80x}^{\$80 \text{ per pair}}$$

5c. Determine the break-even point. Describe what this means.

The system is $\begin{cases} y = 300{,}000 + 30x \\ y = 80x \end{cases}$

The break-even point is where $R(x) = C(x)$.

$R(x) = C(x)$

$80x = 300{,}000 + 30x$

$50x = 300{,}000$

$x = 6000$

Back-substitute to find y: $y = 80x$

$y = 80(6000)$

$y = 480{,}000$

The break-even point is (6000, 480,000).

This means the company will break even when it produces and sells 6000 pairs of shoes. At this level, both revenue and costs are $480,000.

5b. Write the revenue function, R, from the sale of x canoes.

5c. Determine the break-even point. Describe what this means.

Answers for Pencil Problems *(Textbook Exercise references in parentheses)*:

1. The ordered pair is a solution to the system. *(5.1 #1)*

2. $\{(1,3)\}$ *(5.1 #5)*

3. $\{(1,-2)\}$ *(5.1 #27)*

4a. $\varnothing$ or $\{\ \}$ *(5.1 #31)*

4b. $\{(x,y) \mid y = 3x - 5\}$ or $\{(x,y) \mid 21x - 35 = 7y\}$ *(5.1 #33)*

5a. $C(x) = 18{,}000 + 20x$ *(5.1 #61a)* **5b.** $R(x) = 80x$ *(5.1 #61b)*

5c. Break-even point: (300, 24,000). Which means when 300 canoes are produced the company will break-even with cost and revenue at $24,000. *(5.1 #61c)*

Section 5.2
Systems of Linear Equations in Three Variables

Hit the BRAKES!

Did you know that a mathematical model can be used to describe the relationship between the number of feet a car travels once the brakes are applied and the number of seconds the car is in motion after the brakes are applied?

In the Exercise Set of this section, using data collected by a research firm, you will be asked to write the mathematical model that describes this situation.

Objective #1: Verify the solution of a system of linear equations in three variables.

✔ *Solved Problem #1*

1. Show that the ordered triple $(-1,-4,5)$ is a solution of the system:

$$\begin{cases} x - 2y + 3z = 22 \\ 2x - 3y - z = 5 \\ 3x + y - 5z = -32 \end{cases}$$

Test the ordered triple in each equation.

$$x - 2y + 3z = 22$$
$$(-1) - 2(-4) + 3(5) = 22$$
$$22 = 22, \text{ true}$$

$$2x - 3y - z = 5$$
$$2(-1) - 3(-4) - (5) = 5$$
$$5 = 5, \text{ true}$$

$$3x + y - 5z = -32$$
$$3(-1) + (-4) - 5(5) = -32$$
$$-32 = -32, \text{ true}$$

The ordered triple (–1, –4, 5) makes all three equations true, so it is a solution of the system.

Pencil Problem #1

1. Determine if the ordered triple $(2,-1,3)$ is a solution of the system:

$$\begin{cases} x + y + z = 4 \\ x - 2y - z = 1 \\ 2x - y - z = -1 \end{cases}$$

Objective #2: Solve systems of linear equations in three variables.

✔ Solved Problem #2

2. Solve the system: $\begin{cases} x+4y-z=20 \\ 3x+2y+z=8 \\ 2x-3y+2z=-16 \end{cases}$

Add the first two equations to eliminate z.

$$\begin{array}{l} x+4y-z=20 \\ \underline{3x+2y+z=\ 8} \\ 4x+6y \quad\ =28 \end{array}$$

Multiply the first equation by 2 and add it to the third equation to eliminate z again.

$$\begin{array}{l} 2x+8y-2z=40 \\ \underline{2x-3y+2z=-16} \\ 4x+5y \qquad =24 \end{array}$$

Solve the system of two equations in two variables.

$$4x+6y=28$$
$$4x+5y=24$$

Multiply the second equation by -1 and add the equations.

$$\begin{array}{r} 4x+6y=28 \\ \underline{-4x-5y=-24} \\ y=4 \end{array}$$

Back-substitute 4 for y to find x.

$$\begin{aligned} 4x+6y&=28 \\ 4x+6(4)&=28 \\ 4x+24&=28 \\ 4x&=4 \\ x&=1 \end{aligned}$$

Back-substitute into an original equation.

$$\begin{aligned} 3x+2y+z&=8 \\ 3(1)+2(4)+z&=8 \\ 11+z&=8 \\ z&=-3 \end{aligned}$$

The solution is $(1,4,-3)$ and the solution set is $\{(1,4,-3)\}$.

✎ Pencil Problem #2 ✎

2. Solve the system: $\begin{cases} 4x-y+2z=11 \\ x+2y-z=-1 \\ 2x+2y-3z=-1 \end{cases}$

Objective #3: Solve problems using systems in three variables.

✔ *Solved Problem #3*

3. Find the quadratic function $y = ax^2 + bx + c$ whose graph passes through the points $(1,4)$, $(2,1)$, and $(3,4)$.

Use each ordered pair to write an equation.

$(1,4)$: $y = ax^2 + bx + c$

$4 = a(1)^2 + b(1) + c$

$4 = a + b + c$

$(2,1)$: $y = ax^2 + bx + c$

$1 = a(2)^2 + b(2) + c$

$1 = 4a + 2b + c$

$(3,4)$: $y = ax^2 + bx + c$

$4 = a(3)^2 + b(3) + c$

$4 = 9a + 3b + c$

The system of three equations in three variables is:

$$\begin{cases} a + b + c = 4 \\ 4a + 2b + c = 1 \\ 9a + 3b + c = 4 \end{cases}$$

Solve the system: $\begin{cases} a + b + c = 4 \\ 4a + 2b + c = 1 \\ 9a + 3b + c = 4 \end{cases}$

Multiply the first equation by –1 and add it to the second equation:

$$\begin{array}{r} -a - b - c = -4 \\ 4a + 2b + c = 1 \\ \hline 3a + b = -3 \end{array}$$

(Continued on next page)

Pencil Problem #3

3. Find the quadratic function $y = ax^2 + bx + c$ whose graph passes through the points $(-1,6)$, $(1,4)$, and $(2,9)$.

Multiply the first equation by –1 and add it to the third equation:

$$\begin{aligned} -a-b-c &= -4 \\ 9a+3b+c &= 4 \\ \hline 8a+2b \quad &= 0 \end{aligned}$$

Solve this system of two equations in two variables.

$$3a+b=-3$$
$$8a+2b=0$$

Multiply the first equation by –2 and add to the second equation:

$$\begin{aligned} -6a-2b &= 6 \\ 8a+2b &= 0 \\ \hline 2a &= 6 \\ a &= 3 \end{aligned}$$

Back-substitute to find b:

$$\begin{aligned} 3a+b &= -3 \\ 3(3)+b &= -3 \\ 9+b &= -3 \\ b &= -12 \end{aligned}$$

Back-substitute into an original equation to find c:

$$\begin{aligned} a+b+c &= 4 \\ (3)+(-12)+c &= 4 \\ -9+c &= 4 \\ c &= 13 \end{aligned}$$

The quadratic function is $y=3x^2-12x+13$.

Answers for Pencil Problems *(Textbook Exercise references in parentheses)*:

1. Not a solution *(5.2 #1)*

2. $\{(2,-1,1)\}$ *(5.2 #7)*

3. $y=2x^2-x+3$ *(5.2 #19)*

Section 5.3
Partial Fractions

Where's the "UNDO" Button?

We have learned how to write a sum or difference of rational expressions as a single rational expression and saw how this skill is necessary when solving rational inequalities. However, in calculus, it is sometimes necessary to write a single rational expression as a sum or difference of simpler rational expressions, undoing the process of adding or subtracting.
In this section, you will learn how to break up a rational expression into sums or differences of rational expressions with simpler denominators.

Objective #1: Decompose $\frac{P}{Q}$, where Q has only distinct linear factors.

✔ *Solved Problem #1*

1. Find the partial fraction decomposition of $\frac{5x-1}{(x-3)(x+4)}$.

Write a constant over each distinct linear factor in the denominator.

$$\frac{5x-1}{(x-3)(x+4)}=\frac{A}{x-3}+\frac{B}{x+4}$$

Multiply by the LCD, $(x-3)(x+4)$, to eliminate fractions. Then simplify and rearrange terms.

$$(x-3)(x+4)\frac{5x-1}{(x-3)(x+4)}=(x-3)(x+4)\frac{A}{x-3}+(x-3)(x+4)\frac{B}{x+4}$$

$$5x-1=A(x+4)+B(x-3)$$

$$5x-1=Ax+4A+Bx-3B$$

$$5x-1=(A+B)x+(4A-3B)$$

Equating the coefficients of x and equating the constant terms, we obtain a system of equations.

$$\begin{cases} A+B=5 \\ 4A-3B=-1 \end{cases}$$

Multiplying the first equation by 3 and adding it to the second equation, we obtain $7A = 14$, so $A = 2$. Substituting 2 for A in either equation, we obtain $B = 3$. The partial fraction decomposition is

$$\frac{5x-1}{(x-3)(x+4)}=\frac{2}{x-3}+\frac{3}{x+4}.$$

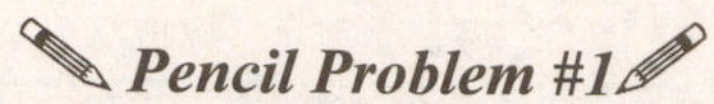

Pencil Problem #1

1. Find the partial fraction decomposition of $\frac{3x+50}{(x-9)(x+2)}$.

Objective #2: Decompose $\frac{P}{Q}$, where Q has repeated linear factors.

✔ Solved Problem #2

2. Find the partial fraction decomposition of $\frac{x+2}{x(x-1)^2}$.

Include one fraction for each power of $x-1$.

$$\frac{x+2}{x(x-1)^2}=\frac{A}{x}+\frac{B}{x-1}+\frac{C}{(x-1)^2}$$

Multiply by the LCD, $x(x-1)^2$, to eliminate fractions. Then simplify and rearrange terms.

$$x(x-1)^2\frac{x+2}{x(x-1)^2}=x(x-1)^2\frac{A}{x}+x(x-1)^2\frac{B}{x-1}+x(x-1)^2\frac{C}{(x-1)^2}$$

$$x+2=A(x-1)^2+Bx(x-1)+Cx$$

$$x+2=A(x^2-2x+1)+Bx(x-1)+Cx$$

$$x+2=Ax^2-2Ax+A+Bx^2-Bx+Cx$$

$$0x^2+x+2=(A+B)x^2+(-2A-B+C)x+A$$

Equating the coefficients of like terms, we obtain a system of equations.

$$\begin{cases} A+B=0 \\ -2A-B+C=1 \\ A=2 \end{cases}$$

We see immediately that $A=2$. Substituting 2 for A in the first equation, we obtain $B=-2$. Substituting these values into the second equation, we obtain $C=3$. The partial fraction decomposition is

$$\frac{x+2}{x(x-1)^2}=\frac{2}{x}+\frac{-2}{x-1}+\frac{3}{(x-1)^2} \text{ or } \frac{2}{x}-\frac{2}{x-1}+\frac{3}{(x-1)^2}.$$

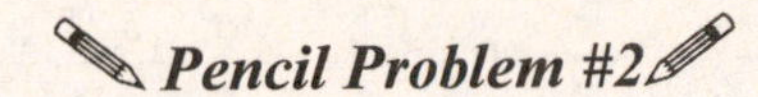

Pencil Problem #2

2. Find the partial fraction decomposition of $\dfrac{x^2}{(x-1)^2(x+1)}$.

Objective #3: Decompose $\dfrac{P}{Q}$, where Q has a nonrepeated prime quadratic factor.

✔ Solved Problem #3

3. Find the partial fraction decomposition of $\dfrac{8x^2+12x-20}{(x+3)(x^2+x+2)}$.

Use a constant over the linear factor and a linear expression over the prime quadratic factor.

$$\frac{8x^2+12x-20}{(x+3)(x^2+x+2)}=\frac{A}{x+3}+\frac{Bx+C}{x^2+x+2}$$

Multiply by the LCD, $(x+3)(x^2+x+2)$, to eliminate fractions. Then simplify and rearrange terms.

$$(x+3)(x^2+x+2)\frac{8x^2+12x-20}{(x+3)(x^2+x+2)}=(x+3)(x^2+x+2)\frac{A}{x+3}+(x+3)(x^2+x+2)\frac{Bx+C}{x^2+x+2}$$

$$8x^2+12x-20=A(x^2+x+2)+(Bx+C)(x+3)$$

$$8x^2+12x-20=Ax^2+Ax+2A+Bx^2+3Bx+Cx+3C$$

$$8x^2+12x-20=(A+B)x^2+(A+3B+C)x+(2A+3C)$$

Equating the coefficients of like terms, we obtain a system of equations.

$$\begin{cases} A+B=8 \\ A+3B+C=12 \\ 2A+3C=-20 \end{cases}$$

Multiply the second equation by −3 and add to the third equation to obtain $-A-9B=-56$. Add this result to the first equation in the system above to obtain $-8B=-48$, so $B=6$. Substituting this value into the first equation, we obtain $A=2$. Substituting the value of A into the third equation, we obtain $C=-8$.

The partial fraction decomposition is

$$\frac{8x^2+12x-20}{(x+3)(x^2+x+2)}=\frac{2}{x+3}+\frac{6x-8}{x^2+x+2}.$$

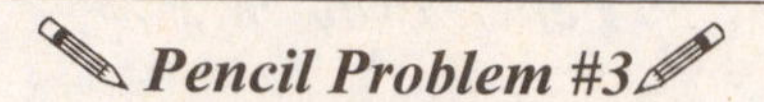

3. Find the partial fraction decomposition of $\dfrac{5x^2+6x+3}{(x+1)(x^2+2x+2)}$.

Objective #4: $\dfrac{P}{Q}$, where Q has a prime, repeated quadratic factor

✔ *Solved Problem #4*

4. Find the partial fraction decomposition of $\dfrac{2x^3+x+3}{(x^2+1)^2}$.

Include one fraction with a linear numerator for each power of x^2+1.

$$\frac{2x^3+x+3}{(x^2+1)^2}=\frac{Ax+B}{x^2+1}+\frac{Cx+D}{(x^2+1)^2}$$

Multiply by the LCD, $(x^2+1)^2$, to eliminate fractions. Then simplify and rearrange terms.

$$(x^2+1)^2\frac{2x^3+x+3}{(x^2+1)^2}=(x^2+1)^2\frac{Ax+B}{x^2+1}+(x^2+1)^2\frac{Cx+D}{(x^2+1)^2}$$

$$2x^3+x+3=(Ax+B)(x^2+1)+Cx+D$$

$$2x^3+x+3=Ax^3+Ax+Bx^2+B+Cx+D$$

$$2x^3+x+3=Ax^3+Bx^2+(A+C)x+(B+D)$$

Equating the coefficients of like terms, we obtain a system of equations.

$$\begin{cases} A=2 \\ B=0 \\ A+C=1 \\ B+D=3 \end{cases}$$

We immediately see that $A=2$ and $B=0$. By performing appropriate substitutions, we obtain $C=-1$ and $D=3$. The partial fraction decomposition is

$$\frac{2x^3+x+3}{(x^2+1)^2}=\frac{2x}{x^2+1}+\frac{-x+3}{(x^2+1)^2}.$$

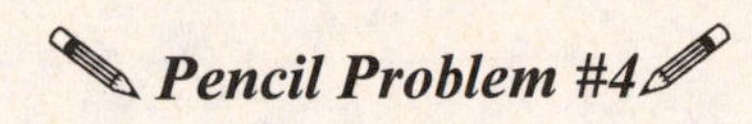

4. Find the partial fraction decomposition of $\dfrac{x^3+x^2+2}{(x^2+2)^2}$.

Answers for Pencil Problems *(Textbook Exercise references in parentheses)*:

1. $\frac{7}{x-9}-\frac{4}{x+2}$ (5.3 #11)
2. $\frac{1}{4(x+1)}+\frac{3}{4(x-1)}+\frac{1}{2(x-1)^2}$ (5.3 #27)
3. $\frac{2}{x+1}+\frac{3x-1}{x^2+2x+2}$ (5.3 #31)
4. $\frac{x+1}{x^2+2}-\frac{2x}{(x^2+2)^2}$ (5.3 #37)

Section 5.4
Systems of Nonlinear Equations in Two Variables

DO YOU FEEL SAFE????

Scientists debate the probability that a "doomsday rock" will collide with Earth. It has been estimated that an asteroid crashes into Earth about once every 250,000 years, and that such a collision would have disastrous results.

Understanding the path of Earth and the path of a comet is essential to detecting threatening space debris. Orbits about the sun are not described by linear equations. The ability to solve systems that do not contain linear equations provides NASA scientists watching for troublesome asteroids with a way to locate possible collision points with Earth's orbit.

Objective #1: Recognize systems of nonlinear equations in two variables.

Solved Problem #1

1a. True or false: A solution of a nonlinear system in two variables is an ordered pair of real numbers that satisfies at least one equation in the system.

False; a solution must satisfy *all* equations in the system.

1b. True or false: The solution of a system of nonlinear equations corresponds to the intersection points of the graphs in the system.

True; each solution will correspond to an intersection point of the graphs.

Pencil Problem #1

1a. True or false: A system of nonlinear equations cannot contain a linear equation.

1b. True or false: The graphs of the equations in a nonlinear system could be a parabola and a circle.

Objective #2: Solve nonlinear systems by substitution.

✔ Solved Problem #2

2. Solve by the substitution method:

$$\begin{cases} x+2y=0 \\ (x-1)^2+(y-1)^2=5 \end{cases}$$

Solve the first equation for x: $x+2y=0$

$$x=-2y$$

Substitute the expression $-2y$ for x in the second equation and solve for y.

$$(x-1)^2+(y-1)^2=5$$

$$(\overbrace{-2y}^{x}-1)^2+(y-1)^2=5$$

$$4y^2+4y+1+y^2-2y+1=5$$

$$5y^2+2y-3=0$$

$$(5y-3)(y+1)=0$$

$$5y-3=0 \quad \text{or} \quad y+1=0$$

$$y=\tfrac{3}{5} \quad \text{or} \quad y=-1$$

If $y=\frac{3}{5}$, $x=-2\left(\frac{3}{5}\right)=-\frac{6}{5}$.
If $y=-1$, $x=-2(-1)=2$.

Check $(2,-1)$ in both original equations.

$$x+2y=0 \qquad (x-1)^2+(y-1)^2=5$$

$$2+2(-1)=0 \qquad (2-1)^2+(-1-1)^2=5$$

$$0=0, \text{ true} \qquad 1+4=5$$

$$5=5, \text{ true}$$

Check $\left(-\frac{6}{5},\frac{3}{5}\right)$ in both original equations.

$$x+2y=0 \qquad (x-1)^2+(y-1)^2=5$$

$$-\tfrac{6}{5}+2\left(\tfrac{3}{5}\right)=0 \qquad \left(-\tfrac{6}{5}-1\right)^2+\left(\tfrac{3}{5}-1\right)^2=5$$

$$-\tfrac{6}{5}+\tfrac{6}{5}=0 \qquad \tfrac{121}{25}+\tfrac{4}{25}=5$$

$$0=0, \text{ true} \qquad \tfrac{125}{25}=5$$

$$5=5, \text{ true}$$

The solution set is $\left\{\left(-\frac{6}{5},\frac{3}{5}\right),(2,-1)\right\}$.

✎ Pencil Problem #2 ✎

2. Solve by the substitution method:

$$\begin{cases} x+y=2 \\ y=x^2-4x+4 \end{cases}$$

Objective #3: Solve nonlinear systems by addition.

✔ Solved Problem #3

3. Solve by the addition method:

$$\begin{cases} y = x^2 + 5 \\ x^2 + y^2 = 25 \end{cases}$$

Arrange the first equation so that variable terms appear on the left, and constants appear on the right.
Add the resulting equations to eliminate the x^2-terms and solve for y.

$$\begin{aligned} -x^2 + y &= 5 \\ x^2 + y^2 &= 25 \\ \hline y^2 + y &= 30 \end{aligned}$$

Solve the resulting quadratic equation.

$$y^2 + y - 30 = 0$$
$$(y+6)(y-5) = 0$$
$$y+6=0 \quad \text{or} \quad y-5=0$$
$$y=-6 \quad \text{or} \quad y=5$$

If $y=-6$,

$$x^2 + (-6)^2 = 25$$
$$x^2 + 36 = 25$$
$$x^2 = -11$$

When $y=-6$ there is no real solution.

If $y=5$,

$$x^2 + (5)^2 = 25$$
$$x^2 + 25 = 25$$
$$x^2 = 0$$
$$x = 0$$

Check $(0,5)$ in both original equations.

$$y = x^2 + 5 \qquad x^2 + y^2 = 25$$
$$5 = (0)^2 + 5 \qquad 0^2 + 5^2 = 25$$
$$5 = 5, \text{ true} \qquad 25 = 25, \text{ true}$$

The solution set is $\{(0,5)\}$.

Pencil Problem #3

3. Solve by the addition method:

$$\begin{cases} x^2 + y^2 = 13 \\ x^2 - y^2 = 5 \end{cases}$$

Objective #4: Solve problems using systems of nonlinear equations.

✔ *Solved Problem #4*

4. Find the length and width of a rectangle whose perimeter is 20 feet and whose area is 21 square feet.

The system is $\begin{cases} 2x+2y=20 \\ xy=21. \end{cases}$

Solve the second equation for x: $xy=21$

$$x=\frac{21}{y}$$

Substitute the expression $\frac{21}{y}$ for x in the first equation and solve for y.

$$2x+2y=20$$
$$2\left(\frac{21}{y}\right)+2y=20$$
$$\frac{42}{y}+2y=20$$
$$42+2y^2=20y$$
$$2y^2-20y+42=0$$
$$y^2-10y+21=0$$
$$(y-7)(y-3)=0$$
$$y-7=0 \quad \text{or} \quad y-3=0$$
$$y=7 \quad \text{or} \quad y=3$$

If $y=7$, $x=\frac{21}{7}=3$.

If $y=3$, $x=\frac{21}{3}=7$.

The dimensions are 7 feet by 3 feet.

Pencil Problem #4

4. The sum of two numbers is 10 and their product is 24. Find the numbers.

Answers for Pencil Problems *(Textbook Exercise references in parentheses)*:

1a. false *(5.4 #1)* **1b.** true *(5.4 #27)* **2.** $\{(1,1),(2,0)\}$ *(5.4 #3)*

3. $\{(-3,-2),(-3,2),(3,-2),(3,2)\}$ *(5.4 #19)* **4.** 6 and 4 *(5.4 #43)*

Section 5.5
Systems of Inequalities

Does Your Weight Fit You?

This chapter opened by noting that the modern emphasis on thinness as the ideal body shape has been suggested as a major cause of eating disorders. In this section, the textbook will demonstrate how systems of linear inequalities in two variables can enable you to establish a healthy weight range for your height and age.

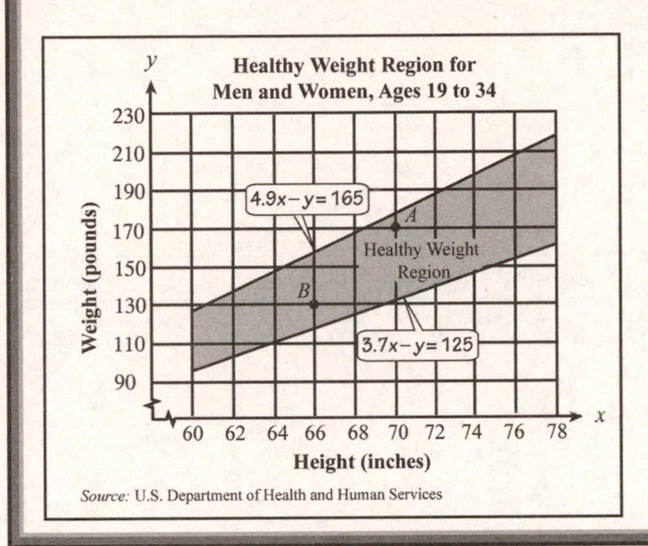

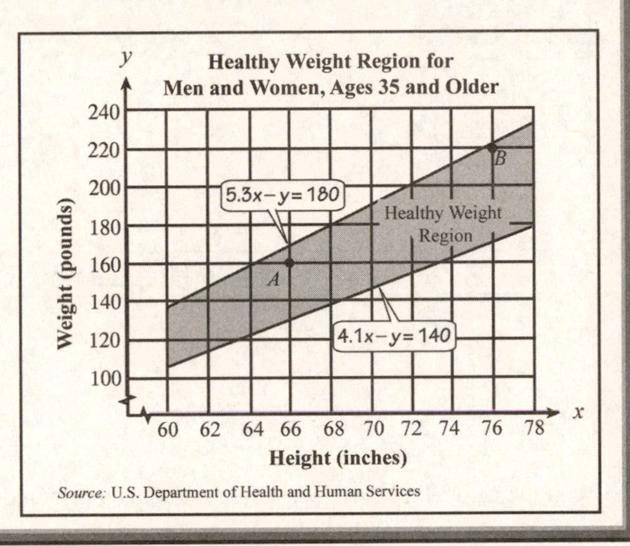

Objective #1: Graph a linear inequality in two variables.

✔ Solved Problem #1

1a. Graph: $4x-2y\geq 8$.

First, graph the equation $4x-2y=8$ with a solid line.

Find the x–intercept:

$$4x-2y=8$$
$$4x-2(0)=8$$
$$4x=8$$
$$x=2$$

Find the y–intercept:

$$4x-2y=8$$
$$4(0)-2y=8$$
$$-2y=8$$
$$y=-4$$

Next, use the origin as a test point.

$$4x-2y\geq 8$$
$$4(0)-2(0)\geq 8$$
$$0\geq 8,\ \text{false}$$

Since the statement is false, shade the half-plane that does not contain the test point.

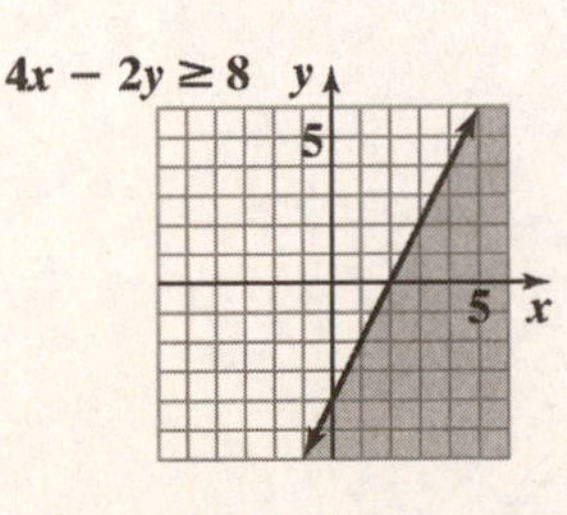

Pencil Problem #1

1a. Graph: $x-2y>10$.

1b. Graph: $y > 1$.

Graph the line $y = 1$ with a dashed line.

Since the inequality is of the form $y > a$, shade the half-plane above the line.

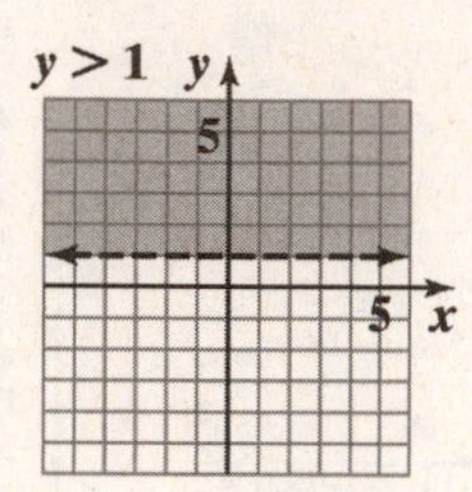

1b. Graph: $x \leq 1$.

Objective #2: Graph a nonlinear inequality in two variables..

✔ *Solved Problem #2*

2. Graph: $x^2 + y^2 \geq 16$.

The graph of $x^2 + y^2 = 16$ is a circle of radius 4 centered at the origin. Use a solid circle because equality is included in ≥.

The point (0, 0) is not on the circle, so we use it as a test point. The result is $0 \geq 16$, which is false. Since the point (0, 0) is inside the circle, the region outside the circle belongs to the solution set. Shade the region outside the circle.

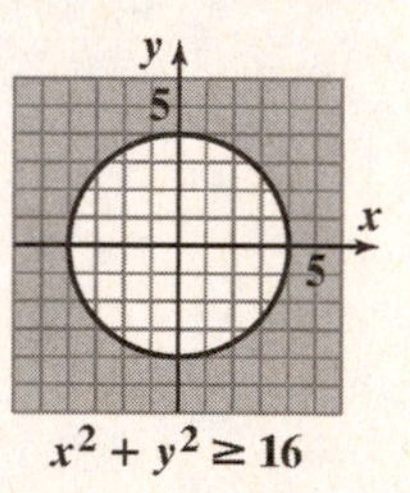

Pencil Problem #2

2. Graph: $x^2 + y^2 > 25$.

Objective #3: Use mathematical models involving systems of linear inequalities.

✔ *Solved Problem #3*

3. The healthy weight region for men and women ages 19 to 34 can be modeled by the following system of linear inequalities:

$$\begin{cases} 4.9x - y \geq 165 \\ 3.7x - y \leq 125 \end{cases}$$

Show that (66, 130) is a solution of the system of inequalities that describes healthy weight for this age group.

Substitute the coordinates of (66, 130) into both inequalities of the system.

$$\begin{cases} 4.9x - y \geq 165 \\ 3.7x - y \leq 125 \end{cases}$$

$$\begin{aligned} 4.9x - y &\geq 165 \\ 4.9(66) - 130 &\geq 165 \\ 193.4 &\geq 165, \text{ true} \end{aligned}$$

$$\begin{aligned} 3.7x - y &\leq 125 \\ 3.7(66) - 130 &\leq 125 \\ 114.2 &\leq 125, \text{ true} \end{aligned}$$

(66, 130) is a solution of the system.

Pencil Problem #3

3. The healthy weight region for men and women ages 35 and older can be modeled by the following system of linear inequalities:

$$\begin{cases} 5.3x - y \geq 180 \\ 4.1x - y \leq 14 \end{cases}$$

Show that (66, 160) is a solution of the system of inequalities that describes healthy weight for this age group.

Objective #4: Graph a system of linear inequalities.

✔ *Solved Problem #4*

4. Graph the solution set of the system:

$$\begin{cases} x - 3y < 6 \\ 2x + 3y \geq -6 \end{cases}$$

Graph the line $x - 3y = 6$ with a dashed line.
Graph the line $2x + 3y = -6$ with a solid line.

For $x - 3y < 6$ use a test point such as (0, 0).

$$\begin{aligned} x - 3y &< 6 \\ 0 - 3(0) &< 6 \\ 0 &< 6, \text{ true} \end{aligned}$$

Since the statement is true, shade the half-plane that contains the test point.

(Continued on next page)

Pencil Problem #4

4. Graph the solution set of the system:

$$\begin{cases} y > 2x - 3 \\ y < -x + 6 \end{cases}$$

For $2x+3y\ge -6$ use a test point such as (0, 0).

$$2x+3y\ge -6$$
$$2(0)+3(0)\ge -6$$
$$0\ge -6, \text{ true}$$

Since the statement is true, shade the half-plane that contains the test point.

The solution set of the system is the intersection (the overlap) of the two half-planes.

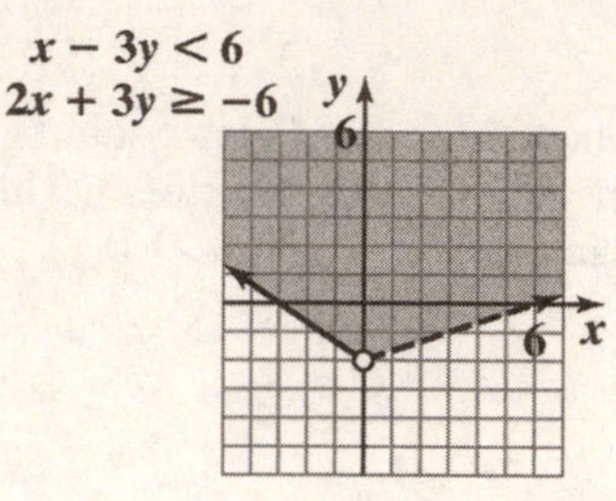

Answers for Pencil Problems *(Textbook Exercise references in parentheses)*:

1a. 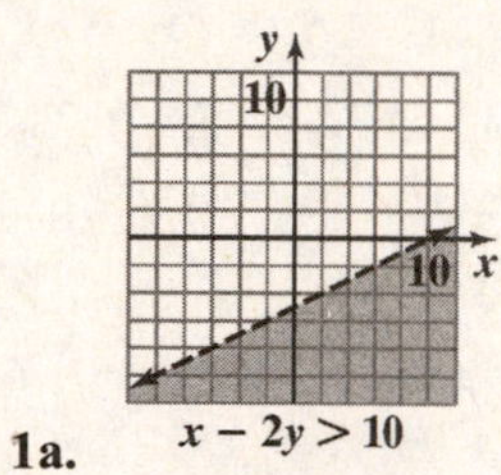

(5.5 #3)

1b. 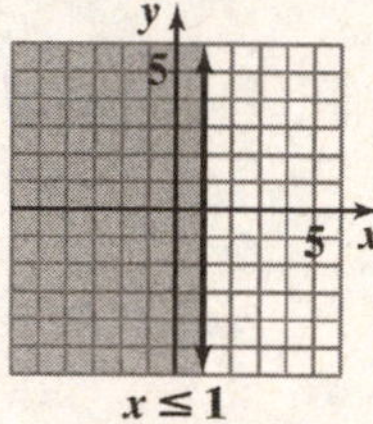

(5.5 #9)

2. 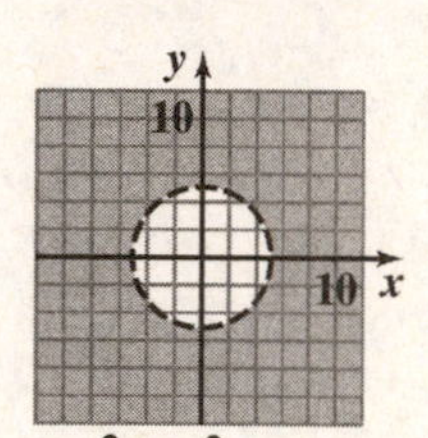

(5.5 #15)

3.

$$5.3x-y\ge 180$$
$$5.3(66)-160\ge 180$$
$$189.8\ge 180, \text{ true}$$

$$4.1x-y\le 14$$
$$4.1(66)-160\le 140$$
$$110.6\le 140, \text{ true}$$

(5.5 #77)

4. 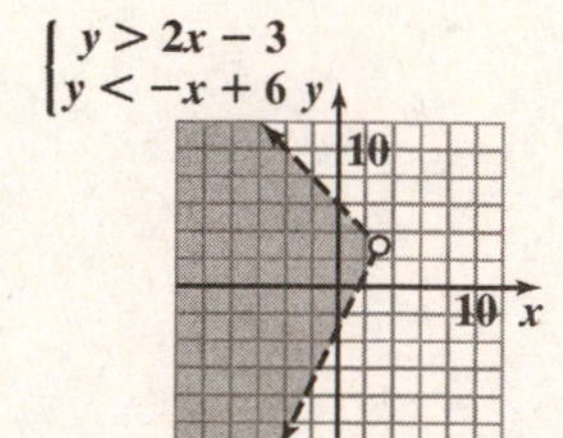

Section 5.6
Linear Programming

MAXIMUM Output with *MINIMUM* Effort!

Many situations in life involve quantities that must be maximized or minimized. Businesses are interested in maximizing profit and minimizing costs.

In the Exercise Set for this section of the textbook, you will encounter a manufacturer looking to maximize profits from selling two models of mountain bikes. But there is a limited amount of time available to assemble and paint these bikes. We will learn how to balance these constraints and determine the proper number of each bike that should be produced.

Objective #1:

Write an objective function describing a quantity that must be maximized or minimized.

✔ *Solved Problem #1*

1. A company manufactures bookshelves and desks for computers. Let x represent the number of bookshelves manufactured daily and y the number of desks manufactured daily. The company's profits are \$25 per bookshelf and \$55 per desk.

Write the objective function that models the company's total daily profit, z, from x bookshelves and y desks.

The total profit is 25 times the number of bookshelves, x, plus 55 times the number of desks, y.

$$\underbrace{z}_{\text{profit}} = \underbrace{25x}_{\text{\$25 for each bookshelf}} + \underbrace{55y}_{\text{\$55 for each desk}}$$

The objective function is $z = 25x + 55y$.

Pencil Problem #1

1. A television manufacturer makes rear-projection and plasma televisions. The profit per unit is \$125 for the rear-projection televisions and \$200 for the plasma televisions. Let x = the number of rear-projection televisions manufactured in a month and y = the number of plasma televisions manufactured in a month.

Write the objective function that models the total monthly profit.

Objective #2: Use inequalities to describe limitations in a situation.

✔ Solved Problem #2

2. Recall that the company in *Solved Problem #1* manufactures bookshelves and desks for computers. x represents the number of bookshelves manufactured daily and y the number of desks manufactured daily.

2a. Write an inequality that models the following constraint: To maintain high quality, the company should not manufacture more than a total of 80 bookshelves and desks per day.

$x+y\le 80$

2b. Write an inequality that models the following constraint: To meet customer demand, the company must manufacture between 30 and 80 bookshelves per day, inclusive.

$30\le x\le 80$

2c. Write an inequality that models the following constraint: The company must manufacture at least 10 and no more than 30 desks per day.

$10\le y\le 30$

2d. Summarize what you have described about this company by writing the objective function for its profits (from *Solved Problem #1*) and the three constraints.

Objective function: $z=25x+55y$.

Constraints: $\begin{cases} x+y\le 80 \\ 30\le x\le 80 \\ 10\le y\le 30 \end{cases}$

Pencil Problem #2

2. Recall that the manufacturer in *Pencil Problem #1* makes rear-projection and plasma televisions. x represents the number of rear-projection televisions manufactured monthly and y the number of plasma televisions manufactured monthly.

2a. Write an inequality that models the following constraint: Equipment in the factory allows for making at most 450 rear-projection televisions in one month.

2b. Write an inequality that models the following constraint: Equipment in the factory allows for making at most 200 plasma televisions in one month.

2c. Write an inequality that models the following constraint: The cost to the manufacturer per unit is $600 for the rear-projection televisions and $900 for the plasma televisions. Total monthly costs cannot exceed $360,000.

2d. Summarize what you have described about this company by writing the objective function for its profits (from *Pencil Problem #1*) and the three constraints.

Objective #3: Use linear programming to solve problems.

✔ Solved Problem #3

3a. For the company in *Solved Problems #1 and 2*, how many bookshelves and how many desks should be manufactured per day to obtain maximum profit? What is the maximum daily profit?

Graph the constraints and find the corners, or vertices, of the region of intersection.

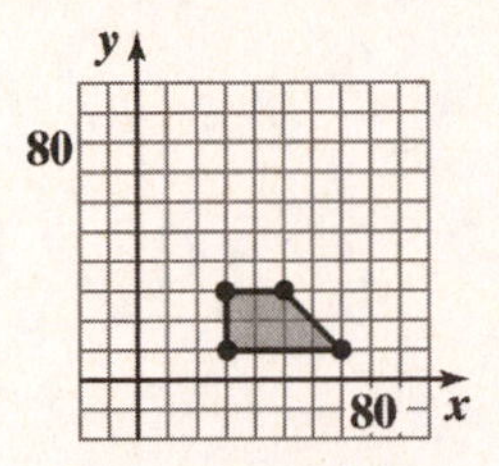

Find the value of the objective function at each corner of the graphed region.

Corner (x, y)	Objective Function $z = 25x + 55y$
$(30,10)$	$z = 25(30) + 55(10)$ $= 750 + 550$ $= 1300$
$(30,30)$	$z = 25(30) + 55(30)$ $= 750 + 1650$ $= 2400$
$(50,30)$	$z = 25(50) + 55(30)$ $= 1250 + 1650$ $= 2900$ (Maximum)
$(70,10)$	$z = 25(70) + 55(10)$ $= 1750 + 550$ $= 2300$

The maximum value of z is 2900 and it occurs at the point (50, 30).

In order to maximize profit, 50 bookshelves and 30 desks must be produced each day for a profit of $2900.

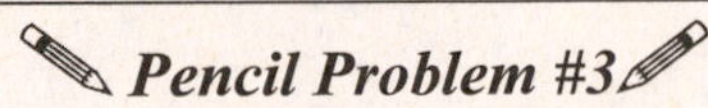

Pencil Problem #3

3a. For the company in *Pencil Problems #1 and 2*, how many rear-projection and plasma televisions should be manufactured per month to obtain maximum profit? What is the maximum monthly profit?

3b. Find the maximum value of the objective function $z = 3x + 5y$ subject to the constraints:

$$\begin{cases} x \geq 0, \quad y \geq 0 \\ x + y \geq 1 \\ x + y \leq 6 \end{cases}$$

Graph the region that represents the intersection of the constraints:

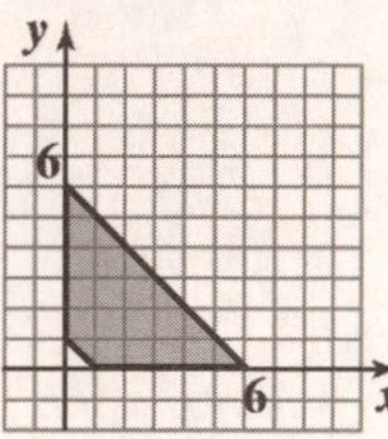

Find the value of the objective function at each corner of the graphed region.

Corner (x, y)	Objective Function $z = 3x + 5y$
$(0,1)$	$z = 3(0) + 5(1) = 5$
$(1,0)$	$z = 3(1) + 5(0) = 3$
$(0,6)$	$z = 3(0) + 5(6) = 30$ (Maximum)
$(6,0)$	$z = 3(6) + 5(0) = 18$

The maximum value is 30.

3b. Find the maximum value of the objective function $z = 4x + y$ subject to the constraints:

$$\begin{cases} x \geq 0, \quad y \geq 0 \\ 2x + 3y \leq 12 \\ x + y \geq 3 \end{cases}$$

Answers for Pencil Problems *(Textbook Exercise references in parentheses)*:

1. $z = 125x + 200y$ *(5.6 #15a)* **2a.** $x \leq 450$ *(5.6 #15b)* **2b.** $y \leq 200$ *(5.6 #15b)*

2c. $600x + 900y \leq 360{,}000$ *(5.6 #15b)* **2d.** $z = 125x + 200y$; $\begin{cases} x \leq 450 \\ y \leq 200 \\ 600x + 900y \leq 360{,}000 \end{cases}$ *(5.6 #15b)*

3a. 300 rear-projection and 200 plasma televisions; Maximum profit: \$77,500 *(5.6 #15e)* **3b.** 24 *(5.6 #7)*

Section 6.1
Matrix Solutions to Linear Systems

Avoiding the Scale?

We have already studied guidelines, given as a set of linear inequalities, that tell how much we should weigh based on age and height. But what do people really weigh? The chart below shows average weights of American adults by age and gender.

In this section of the textbook, we will look at how such data can be placed in something called a matrix and how matrices can be used to solve linear systems.

	Ages 20–29	Ages 30–39	Ages 40–49	Ages 50–59	Ages 60–69	Ages 70–79	Ages 80+
Men	188	194	202	199	198	187	168
Women	156	165	171	172	171	156	142

Source: National Center for Health Statistics

Objective #1: Write the augmented matrix for a linear system.

✔ Solved Problem #1

1. Write the augmented matrix for the system:

$$\begin{cases} 3x + 4y = 19 \\ 2y + 3z = 8 \\ 4x - 5z = 7 \end{cases}$$

The augmented matrix has a row for each equation and a vertical bar separating the coefficients of the variables on the left from the constants on the right. Coefficients of the same variable are lined up vertically in the same column. If a variable is missing from an equation, its coefficient is 0.

It may be helpful to view the system as

$$\begin{cases} 3x + 4y + 0z = 19 \\ 0x + 2y + 3z = 8 \\ 4x + 0y - 5z = 7 \end{cases}$$

The augmented matrix is

$$\left[\begin{array}{ccc|c} 3 & 4 & 0 & 19 \\ 0 & 2 & 3 & 8 \\ 4 & 0 & -5 & 7 \end{array}\right]$$

Pencil Problem #1

1. Write the augmented matrix for the system:

$$\begin{cases} 5x - 2y - 3z = 0 \\ x + y = 5 \\ 2x - 3z = 4 \end{cases}$$

Objective #2: Perform matrix row operations.

✔ *Solved Problem #2*

2a. Perform the row operation and write the new matrix:

$$\left[\begin{array}{ccc|c} 4 & 12 & -20 & 8 \\ 1 & 6 & -3 & 7 \\ -3 & -2 & 1 & -9 \end{array}\right];\ \frac{1}{4}R_1$$

Multiply each element in row 1 by $\frac{1}{4}$. The elements in row 2 and row 3 do not change.

$$\left[\begin{array}{ccc|c} \frac{1}{4}(4) & \frac{1}{4}(12) & \frac{1}{4}(-20) & \frac{1}{4}(8) \\ 1 & 6 & -3 & 7 \\ -3 & -2 & 1 & -9 \end{array}\right] = \left[\begin{array}{ccc|c} 1 & 3 & -5 & 2 \\ 1 & 6 & -3 & 7 \\ -3 & -2 & 1 & -9 \end{array}\right]$$

2b. Perform the row operation and write the new matrix:

$$\left[\begin{array}{ccc|c} 4 & 12 & -20 & 8 \\ 1 & 6 & -3 & 7 \\ -3 & -2 & 1 & -9 \end{array}\right];\ 3R_2 + R_3$$

Multiply each element in row 2 by 3 and add to the corresponding element in row 3. Replace the elements in row 3. Row 1 and row 2 do not change.

$$\left[\begin{array}{ccc|c} 4 & 12 & -20 & 8 \\ 1 & 6 & -3 & 7 \\ 3(1)+(-3) & 3(6)+(-2) & 3(-3)+1 & 3(7)+(-9) \end{array}\right]$$

$$= \left[\begin{array}{ccc|c} 4 & 12 & -20 & 8 \\ 1 & 6 & -3 & 7 \\ 0 & 16 & -8 & 12 \end{array}\right]$$

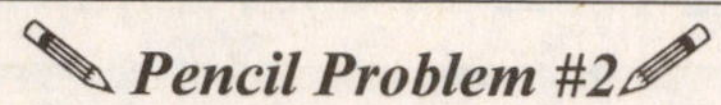

Pencil Problem #2

2a. Perform the row operation and write the new matrix:

$$\left[\begin{array}{ccc|c} 2 & -6 & 4 & 10 \\ 1 & 5 & -5 & 0 \\ 3 & 0 & 4 & 7 \end{array}\right];\ \frac{1}{2}R_1$$

2b. Perform the row operation and write the new matrix:

$$\left[\begin{array}{ccc|c} 1 & -3 & 2 & 0 \\ 3 & 1 & -1 & 7 \\ 2 & -2 & 1 & 3 \end{array}\right];\ -3R_1 + R_2$$

Objective #3: Use matrices and Gaussian elimination to solve systems.

✔ Solved Problem #3

3. Use matrices to solve the system: $\begin{cases} 2x+y+2z=18 \\ x-y+2z=9 \\ x+2y-z=6 \end{cases}$

Write the augmented matrix for the system.

$$\left[\begin{array}{ccc|c} 2 & 1 & 2 & 18 \\ 1 & -1 & 2 & 9 \\ 1 & 2 & -1 & 6 \end{array}\right]$$

We want a 1 in the upper left position. One way to do this is to interchange row 1 and row 2.

$$\left[\begin{array}{ccc|c} 2 & 1 & 2 & 18 \\ 1 & -1 & 2 & 9 \\ 1 & 2 & -1 & 6 \end{array}\right] R_1 \leftrightarrow R_2 = \left[\begin{array}{ccc|c} 1 & -1 & 2 & 9 \\ 2 & 1 & 2 & 18 \\ 1 & 2 & -1 & 6 \end{array}\right]$$

Now we want zeros below the 1 in the first column.

$$\left[\begin{array}{ccc|c} 1 & -1 & 2 & 9 \\ 2 & 1 & 2 & 18 \\ 1 & 2 & -1 & 6 \end{array}\right] -2R_1 + R_2 = \left[\begin{array}{ccc|c} 1 & -1 & 2 & 9 \\ 0 & 3 & -2 & 0 \\ 1 & 2 & -1 & 6 \end{array}\right]$$

$$\left[\begin{array}{ccc|c} 1 & -1 & 2 & 9 \\ 0 & 3 & -2 & 0 \\ 1 & 2 & -1 & 6 \end{array}\right] -R_1 + R_3 = \left[\begin{array}{ccc|c} 1 & -1 & 2 & 9 \\ 0 & 3 & -2 & 0 \\ 0 & 3 & -3 & -3 \end{array}\right]$$

Next we want a 1 in the second row, second column.

$$\left[\begin{array}{ccc|c} 1 & -1 & 2 & 9 \\ 0 & 3 & -2 & 0 \\ 0 & 3 & -3 & -3 \end{array}\right] \frac{1}{3}R_2 = \left[\begin{array}{ccc|c} 1 & -1 & 2 & 9 \\ 0 & 1 & -\frac{2}{3} & 0 \\ 0 & 3 & -3 & -3 \end{array}\right]$$

Now we want a zero below the 1 in the second row, second column.

$$\left[\begin{array}{ccc|c} 1 & -1 & 2 & 9 \\ 0 & 1 & -\frac{2}{3} & 0 \\ 0 & 3 & -3 & -3 \end{array}\right] -3R_2 + R_3 = \left[\begin{array}{ccc|c} 1 & -1 & 2 & 9 \\ 0 & 1 & -\frac{2}{3} & 0 \\ 0 & 0 & -1 & -3 \end{array}\right]$$

(Continued on next page)

✎ Pencil Problem #3 ✎

3. Use matrices to solve: $\begin{cases} x+y-z=-2 \\ 2x-y+z=5 \\ -x+2y+2z=1 \end{cases}$

Next we want a 1 in the third row, third column.

$$\left[\begin{array}{ccc|c} 1 & -1 & 2 & 9 \\ 0 & 1 & -\frac{2}{3} & 0 \\ 0 & 0 & -1 & -3 \end{array}\right] \quad -R_3 = \left[\begin{array}{ccc|c} 1 & -1 & 2 & 9 \\ 0 & 1 & -\frac{2}{3} & 0 \\ 0 & 0 & 1 & 3 \end{array}\right]$$

The resulting system is:

$$\begin{aligned} x - y + 2z &= 9 \\ y - \frac{2}{3}z &= 0 \\ z &= 3 \end{aligned}$$

Back-substitute 3 for z in the second equation.

$$\begin{aligned} y - \frac{2}{3}(3) &= 0 \\ y - 2 &= 0 \\ y &= 2 \end{aligned}$$

Back-substitute 2 for y and 3 for z in the first equation.

$$\begin{aligned} x - y + 2z &= 9 \\ x - (2) + 2(3) &= 9 \\ x - 2 + 6 &= 9 \\ x + 4 &= 9 \\ x &= 5 \end{aligned}$$

$(5,2,3)$ satisfies both equations.

The solution set is $\{(5,2,3)\}$.

Objective #4: Use matrices and Gauss-Jordan elimination to solve systems.

✔ Solved Problem #4

Solve the system by Gauss-Jordan elimination. Begin with the matrix obtained in Solved Problem #3.

$$\begin{cases} 2x + y + 2z = 18 \\ x - y + 2z = 9 \\ x + 2y - z = 6 \end{cases}$$

The final matrix from Solved Problem #3 is

$$\left[\begin{array}{ccc|c} 1 & -1 & 2 & 9 \\ 0 & 1 & -\frac{2}{3} & 0 \\ 0 & 0 & 1 & 3 \end{array}\right]$$

First we want a zero above the 1 in the second column.

$$\left[\begin{array}{ccc|c} 1 & -1 & 2 & 9 \\ 0 & 1 & -\frac{2}{3} & 0 \\ 0 & 0 & 1 & 3 \end{array}\right] \quad R_2 + R_1 = \left[\begin{array}{ccc|c} 1 & 0 & \frac{4}{3} & 9 \\ 0 & 1 & -\frac{2}{3} & 0 \\ 0 & 0 & 1 & 3 \end{array}\right]$$

Now we want zeros above the 1 in the third column.

$$\left[\begin{array}{ccc|c} 1 & 0 & \frac{4}{3} & 9 \\ 0 & 1 & -\frac{2}{3} & 0 \\ 0 & 0 & 1 & 3 \end{array}\right] \quad -\frac{4}{3}R_3 + R_1 = \left[\begin{array}{ccc|c} 1 & 0 & 0 & 5 \\ 0 & 1 & -\frac{2}{3} & 0 \\ 0 & 0 & 1 & 3 \end{array}\right]$$

$$\left[\begin{array}{ccc|c} 1 & 0 & 0 & 5 \\ 0 & 1 & -\frac{2}{3} & 0 \\ 0 & 0 & 1 & 3 \end{array}\right] \quad \frac{2}{3}R_3 + R_2 = \left[\begin{array}{ccc|c} 1 & 0 & 0 & 5 \\ 0 & 1 & 0 & 2 \\ 0 & 0 & 1 & 3 \end{array}\right]$$

This last matrix corresponds to

$$x = 5, y = 2, z = 3$$

The solution set is $\{(5, 2, 3)\}$.

Pencil Problem #4

4a. Solve the system by Gauss-Jordan elimination. Begin with the matrix obtained in Pencil Problem #3.

$$\begin{cases} x + y - z = -2 \\ 2x - y + z = 5 \\ -x + 2y + 2z = 1 \end{cases}$$

Answers for Pencil Problems *(Textbook Exercise references in parentheses)*:

1. $\left[\begin{array}{ccc|c} 5 & -2 & -3 & 0 \\ 1 & 1 & 0 & 5 \\ 2 & 0 & -3 & 4 \end{array}\right]$ *(6.1 #5)*

2a. $\left[\begin{array}{ccc|c} 1 & -3 & 2 & 5 \\ 1 & 5 & -5 & 0 \\ 3 & 0 & 4 & 7 \end{array}\right]$ *(6.1 #13)* **2b.** $\left[\begin{array}{ccc|c} 1 & -3 & 2 & 0 \\ 0 & 10 & -7 & 7 \\ 2 & -2 & 1 & 3 \end{array}\right]$ *(6.1 #15)*

3. $\{(1, -1, 2)\}$ *(6.1 #21)*

4. $\{(1, -1, 2)\}$ *(6.1 #21)*

Section 6.2
Inconsistent and Dependent Systems and Their Applications

Lane Closed Ahead! Be Prepared to Stop!

You've allowed yourself barely enough time to get to campus, and now you see a sign for road construction along your normal route. Should you take an alternate route to make it to class on time?
In this section of the textbook, we use systems of equations to model traffic flow. Systems of equations with more than one solution can tell us how many cars should be directed toward alternate routes when flow along one street is limited by road work.

Objective #1: Apply Gaussian elimination to systems without unique solutions.

✔ *Solved Problem #1*

1a. Use Gaussian elimination to solve the system:

$$\begin{cases} x - 2y - z = -5 \\ 2x - 3y - z = 0 \\ 3x - 4y - z = 1 \end{cases}$$

Write the augmented matrix for the system.

$$\left[\begin{array}{ccc|c} 1 & -2 & -1 & -5 \\ 2 & -3 & -1 & 0 \\ 3 & -4 & -1 & 1 \end{array}\right]$$

Attempt to simplify the matrix to row-echelon form. The matrix already has a 1 in the upper left position. We want 0s below the 1. Multiply row 1 by –2 and add to row 2, and multiply row 1 by –3 and add to row 3.

$$\left[\begin{array}{ccc|c} 1 & -2 & -1 & -5 \\ 0 & 1 & 1 & 10 \\ 0 & 2 & 2 & 16 \end{array}\right]$$

Now we want a 0 below the 1 in the second column. Multiply row 2 by –2 and add to row 3.

$$\left[\begin{array}{ccc|c} 1 & -2 & -1 & -5 \\ 0 & 1 & 1 & 10 \\ 0 & 0 & 0 & -4 \end{array}\right]$$

The last row represents $0x + 0y + 0z = -4$, which is false. The system has no solution. The solution set is $\varnothing$, the empty set.

Pencil Problem #1

1a. Use Gaussian elimination to solve the system:

$$\begin{cases} 5x + 12y + z = 10 \\ 2x + 5y + 2z = -1 \\ x + 2y - 3z = 5 \end{cases}$$

1b. Use Gaussian elimination to solve the system:

$$\begin{cases} x - 2y - z = 5 \\ 2x - 5y + 3z = 6 \\ x - 3y + 4z = 1 \end{cases}$$

We begin with the augmented matrix.

$$\left[\begin{array}{ccc|c} 1 & -2 & -1 & 5 \\ 2 & -5 & 3 & 6 \\ 1 & -3 & 4 & 1 \end{array}\right]$$

$$\xrightarrow[-R_1+R_3]{-2R_1+R_2} \left[\begin{array}{ccc|c} 1 & -2 & -1 & 5 \\ 0 & -1 & 5 & -4 \\ 0 & -1 & 5 & -4 \end{array}\right]$$

$$\xrightarrow{-R_2} \left[\begin{array}{ccc|c} 1 & -2 & -1 & 5 \\ 0 & 1 & -5 & 4 \\ 0 & -1 & 5 & -4 \end{array}\right]$$

$$\xrightarrow{R_2+R_3} \left[\begin{array}{ccc|c} 1 & -2 & -1 & 5 \\ 0 & 1 & -5 & 4 \\ 0 & 0 & 0 & 0 \end{array}\right]$$

The original system is equivalent to the system

$$\begin{cases} x - 2y - z = 5 \\ y - 5z = 4 \end{cases}$$

The system is consistent and the equations are dependent. Express x and y in terms of z.

$$y - 5z = 4$$
$$y = 5z + 4$$

$$x - 2y - z = 5$$
$$x = 2y + z + 5$$
$$x = 2(5z + 4) + z + 5$$
$$x = 10z + 8 + z + 5$$
$$x = 11z + 13$$

Each ordered pair of the form $(11z + 13, 5z + 4, z)$ is a solution of the system. The solution set is $\{(11z + 13, 5z + 4, z)\}$.

1b. Use Gaussian elimination to solve the system:

$$\begin{cases} 8x + 5y + 11z = 30 \\ -x - 4y + 2z = 3 \\ 2x - y + 5z = 12 \end{cases}$$

Objective #2: Apply Gaussian elimination to systems with more variables than equations.

✔ *Solved Problem #2*

2. Use Gaussian elimination to solve the system:

$$\begin{cases} x + 2y + 3z = 70 \\ x + y + z = 60 \end{cases}$$

We begin with the augmented matrix.

$$\left[\begin{array}{ccc|c} 1 & 2 & 3 & 70 \\ 1 & 1 & 1 & 60 \end{array}\right]$$

$$\xrightarrow{-R_1+R_2} \left[\begin{array}{ccc|c} 1 & 2 & 3 & 70 \\ 0 & -1 & -2 & -10 \end{array}\right]$$

$$\xrightarrow{-R_2} \left[\begin{array}{ccc|c} 1 & 2 & 3 & 70 \\ 0 & 1 & 2 & 10 \end{array}\right]$$

The original system is equivalent to the system

$$\begin{cases} x + 2y + 3z = 70 \\ \quad\;\; y + 2z = 10 \end{cases}$$

Express x and y in terms of z.

$$y + 2z = 10$$
$$y = -2z + 10$$
$$x + 2y + 3z = 70$$
$$x = -2y - 3z + 70$$
$$x = -2(-2z + 10) - 3z + 70$$
$$x = 4z - 20 - 3z + 70$$
$$x = z + 50$$

Each ordered pair of the form $(z + 50, -2z + 10, z)$ is a solution of the system. The solution set is $\{(z + 50, -2z + 10, z)\}$.

Pencil Problem #2

2. Use Gaussian elimination to solve the system:

$$\begin{cases} 2x + y - z = 2 \\ 3x + 3y - 2z = 3 \end{cases}$$

Objective #3: Solve problems involving systems without unique solutions.

✔ Solved Problem #3

3. The figure shows a system of four one-way streets. The numbers in the figure denote the number of cars per minute that travel in the direction shown.

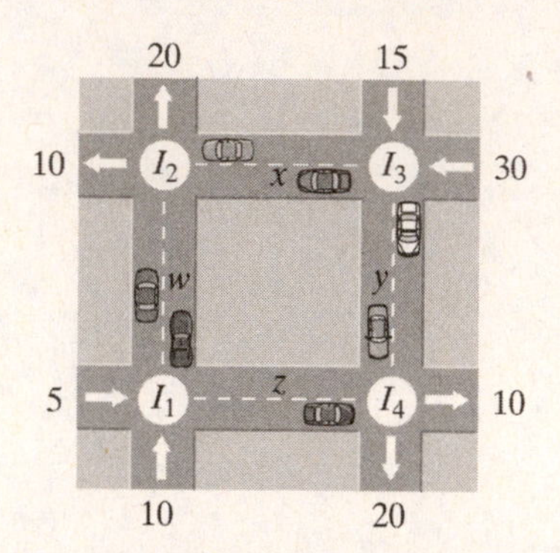

3a. Use the requirement that the number of cars entering each of the intersections per minute must equal the number of cars leaving per minute to set up a system of equations that keeps traffic moving.

Consider one intersection at a time.

I_1: $10 + 5 = 15$ cars enter and $w + z$ leave, so $w + z = 15$.

I_2: $w + x$ cars enter and $10 + 20 = 30$ leave, so $w + x = 30$.

I_3: $15 + 30 = 45$ cars enter and $x + y$ leave, so $x + y = 45$.

I_4: $y + z$ cars enter and $20 + 10 = 30$ leave, so $y + z = 30$.

The system is $\begin{cases} w+z=15 \\ w+x=30 \\ x+y=45 \\ y+z=30 \end{cases}$

Pencil Problem #3

3. The figure shows a system of four one-way streets. The numbers in the figure denote the number of cars per hour that travel in the direction shown.

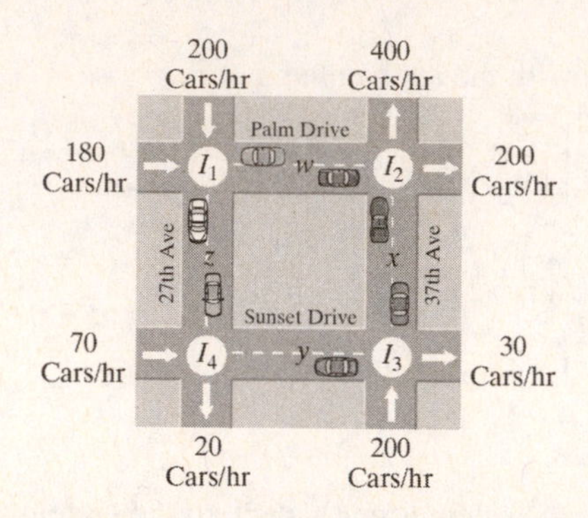

3a. Use the requirement that the number of cars entering each of the intersections per hour must equal the number of cars leaving per hour to set up a system of equations that keeps traffic moving.

3b. Use Gaussian elimination to solve the system.

We begin with the augmented matrix.

$$\left[\begin{array}{cccc|c} 1 & 0 & 0 & 1 & 15 \\ 1 & 1 & 0 & 0 & 30 \\ 0 & 1 & 1 & 0 & 45 \\ 0 & 0 & 1 & 1 & 30 \end{array}\right]$$

$$\xrightarrow{-R_1+R_2} \left[\begin{array}{cccc|c} 1 & 0 & 0 & 1 & 15 \\ 0 & 1 & 0 & -1 & 15 \\ 0 & 1 & 1 & 0 & 45 \\ 0 & 0 & 1 & 1 & 30 \end{array}\right]$$

$$\xrightarrow{-R_2+R_3} \left[\begin{array}{cccc|c} 1 & 0 & 0 & 1 & 15 \\ 0 & 1 & 0 & -1 & 15 \\ 0 & 0 & 1 & 1 & 30 \\ 0 & 0 & 1 & 1 & 30 \end{array}\right]$$

$$\xrightarrow{-R_3+R_4} \left[\begin{array}{cccc|c} 1 & 0 & 0 & 1 & 15 \\ 0 & 1 & 0 & -1 & 15 \\ 0 & 0 & 1 & 1 & 30 \\ 0 & 0 & 0 & 0 & 0 \end{array}\right]$$

From the first row, we get $w + z = 15$, so $w = 15 - z$. From the second and third rows, we get $x - z = 15$ and $y + z = 30$, respectively, so $x = 15 + z$ and $y = 30 - z$. The solution set is $\{(15 - z, 15 + z, 30 - z, z)\}$.

3b. Use Gaussian elimination to solve the system.

3c. If construction limits z to 10 cars per minute, how many cars per minute must pass between the other intersections to keep traffic flowing?

Substitute 10 for z in the system's solution.

$(15 - z, 15 + z, 30 - z, z)$
$= (15 - 10, 15 + 10, 30 - 10, 10)$
$= (5, 25, 20, 10)$

To keep traffic flowing, we must have $w = 5$, $x = 25$, and $y = 20$ cars per minute.

3c. If construction limits z to 50 cars per hour, how many cars per hour must pass between the other intersections to keep traffic flowing?

Answers for Pencil Problems *(Textbook Exercise references in parentheses)*:

1a. No solution or $\varnothing$ *(6.2 #1)* **1b.** $\{(5-2z, -2+z, z)\}$ *(6.2 #7)*

2. $\left\{\left(1+\frac{1}{3}z, \frac{1}{3}z, z\right)\right\}$ *(6.2 #15)*

3a. $\begin{cases} w+z=380 \\ w+x=600 \\ x-y=170 \\ y-z=50 \end{cases}$ *(6.2 #33a)* **3b.** $\{(380-z, 220+z, 50+z, z)\}$ *(6.2 #33b)*

3c. $w=330, x=270, y=100$ *(6.2 #33c)*

Section 6.3
Matrix Operations and Their Applications

Making Things Clearer

Have you ever had trouble reading a document because the text didn't differ sufficiently from the background? By increasing the contrast between the text and the background, you can often make the document easier to read.

In this section of the textbook, we use matrix operations to change the contrast between a letter and its background and to transform figures through translations, stretching or shrinking, and reflections.

Objective #1: Use matrix notation.

✔ *Solved Problem #1*

1. Let $A = \begin{bmatrix} 5 & -2 \\ -3 & \pi \\ 1 & 6 \end{bmatrix}$.

1a. What is the order of A?
The matrix has 3 rows and 2 columns, so it is of order 3×2.

1b. Identify a_{12} and a_{31}.

The element a_{12} is in the first row and second column: $a_{12} = -2$.

The element a_{31} is in the third row and first column: $a_{31} = 1$.

Pencil Problem #1

1. Let $A = \begin{bmatrix} 1 & -5 & \pi & e \\ 0 & 7 & -6 & -\pi \\ -2 & \frac{1}{2} & 11 & -\frac{1}{5} \end{bmatrix}$.

1a. What is the order of A?

1b. Identify a_{32} and a_{23}.

Objective #2: Understand what is meant by equal matrices.

✔ *Solved Problem #2*

2. Find values for the variables so that the matrices are equal.

$$\begin{bmatrix} x & y+1 \\ z & 6 \end{bmatrix} = \begin{bmatrix} 1 & 5 \\ 3 & 6 \end{bmatrix}$$

These matrices are of the same order, so they are equal if and only if corresponding elements are equal.

$x = 1$
$y + 1 = 5$, so $y = 4$
$z = 3$

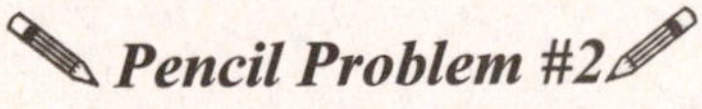

Pencil Problem #2

2. Find values for the variables so that the matrices are equal.

$$\begin{bmatrix} x & 2y \\ z & 9 \end{bmatrix} = \begin{bmatrix} 4 & 12 \\ 3 & 9 \end{bmatrix}$$

Objective #3: Add and subtract matrices.

✔ *Solved Problem #3*

3. Perform the indicated matrix operations.

3a. $\begin{bmatrix} -4 & 3 \\ 7 & -6 \end{bmatrix} + \begin{bmatrix} 6 & -3 \\ 2 & -4 \end{bmatrix}$

Add corresponding elements.

$$\begin{bmatrix} -4+6 & 3+(-3) \\ 7+2 & -6+(-4) \end{bmatrix} = \begin{bmatrix} 2 & 0 \\ 9 & -10 \end{bmatrix}$$

3b. $\begin{bmatrix} 5 & 4 \\ -3 & 7 \\ 0 & 1 \end{bmatrix} - \begin{bmatrix} -4 & 8 \\ 6 & 0 \\ -5 & 3 \end{bmatrix}$

Subtract corresponding elements.

$$\begin{bmatrix} 5-(-4) & 4-8 \\ -3-6 & 7-0 \\ 0-(-5) & 1-3 \end{bmatrix} = \begin{bmatrix} 9 & -4 \\ -9 & 7 \\ 5 & -2 \end{bmatrix}$$

Pencil Problem #3

3. Perform the indicated matrix operations.

3a. $\begin{bmatrix} 1 & 3 \\ 3 & 4 \\ 5 & 6 \end{bmatrix} + \begin{bmatrix} 2 & -1 \\ 3 & -2 \\ 0 & 1 \end{bmatrix}$

3b. $\begin{bmatrix} 4 & 1 \\ 3 & 2 \end{bmatrix} - \begin{bmatrix} 5 & 9 \\ 0 & 7 \end{bmatrix}$

Objective #4: Perform scalar multiplication.

✔ *Solved Problem #4*

4. If $A = \begin{bmatrix} -4 & 1 \\ 3 & 0 \end{bmatrix}$ and $B = \begin{bmatrix} -1 & -2 \\ 8 & 5 \end{bmatrix}$, find each of the following.

4a. $-6B$

$$-6B = -6\begin{bmatrix} -1 & -2 \\ 8 & 5 \end{bmatrix}$$

$$= \begin{bmatrix} -6(-1) & -6(-2) \\ -6(8) & -6(5) \end{bmatrix}$$

$$= \begin{bmatrix} 6 & 12 \\ -48 & -30 \end{bmatrix}$$

Pencil Problem #4

4. If $A = \begin{bmatrix} 2 \\ -4 \\ 1 \end{bmatrix}$ and $B = \begin{bmatrix} -5 \\ 3 \\ -1 \end{bmatrix}$, find each of the following.

4a. $-4A$

4b. $3A + 2B$

$$3A+2B = 3\begin{bmatrix} -4 & 1 \\ 3 & 0 \end{bmatrix} + 2\begin{bmatrix} -1 & -2 \\ 8 & 5 \end{bmatrix}$$
$$= \begin{bmatrix} 3(-4) & 3(1) \\ 3(3) & 3(0) \end{bmatrix} + \begin{bmatrix} 2(-1) & 2(-2) \\ 2(8) & 2(5) \end{bmatrix}$$
$$= \begin{bmatrix} -12 & 3 \\ 9 & 0 \end{bmatrix} + \begin{bmatrix} -2 & -4 \\ 16 & 10 \end{bmatrix}$$
$$= \begin{bmatrix} -12+(-2) & 3+(-4) \\ 9+16 & 0+10 \end{bmatrix}$$
$$= \begin{bmatrix} -14 & -1 \\ 25 & 10 \end{bmatrix}$$

4b. $3A + 2B$

Objective #5: Solve matrix equations.

✔ *Solved Problem #5*

5. Solve for X in the matrix equation $3X + A = B$ where $A = \begin{bmatrix} 2 & -8 \\ 0 & 4 \end{bmatrix}$ and $B = \begin{bmatrix} -10 & 1 \\ -9 & 17 \end{bmatrix}$.

Begin by solving the matrix equation for X.

$$3X + A = B$$
$$3X = B - A$$
$$X = \frac{1}{3}(B - A)$$

Now use matrices A and B to find X.

$$X = \frac{1}{3}\left(\begin{bmatrix} -10 & 1 \\ -9 & 17 \end{bmatrix} - \begin{bmatrix} 2 & -8 \\ 0 & 4 \end{bmatrix}\right)$$
$$= \frac{1}{3}\begin{bmatrix} -12 & 9 \\ -9 & 13 \end{bmatrix}$$
$$= \begin{bmatrix} -4 & 3 \\ -3 & \frac{13}{3} \end{bmatrix}$$

Pencil Problem #5

5. Solve for X in the matrix equation $2X + A = B$ where $A = \begin{bmatrix} -3 & -7 \\ 2 & -9 \\ 5 & 0 \end{bmatrix}$ and $B = \begin{bmatrix} -5 & -1 \\ 0 & 0 \\ 3 & -4 \end{bmatrix}$.

Objective #6: Multiply matrices.

✔ *Solved Problem #6*

6a. Find AB, given $A=\begin{bmatrix}1 & 3\\ 2 & 5\end{bmatrix}$ and $B=\begin{bmatrix}4 & 6\\ 1 & 0\end{bmatrix}$.

$$AB=\begin{bmatrix}1 & 3\\ 2 & 5\end{bmatrix}\begin{bmatrix}4 & 6\\ 1 & 0\end{bmatrix}$$
$$=\begin{bmatrix}1(4)+3(1) & 1(6)+3(0)\\ 2(4)+5(1) & 2(6)+5(0)\end{bmatrix}$$
$$=\begin{bmatrix}7 & 6\\ 13 & 12\end{bmatrix}$$

6b. Find the product, if possible.

$$\begin{bmatrix}1 & 3\\ 0 & 2\end{bmatrix}\begin{bmatrix}2 & 3 & -1 & 6\\ 0 & 5 & 4 & 1\end{bmatrix}$$

The number of columns in the first matrix equals the number of rows in the second matrix, so it is possible to find the product.

$$\begin{bmatrix}1 & 3\\ 0 & 2\end{bmatrix}\begin{bmatrix}2 & 3 & -1 & 6\\ 0 & 5 & 4 & 1\end{bmatrix}$$
$$=\begin{bmatrix}1(2)+3(0) & 1(3)+3(5) & 1(-1)+3(4) & 1(6)+3(1)\\ 0(2)+2(0) & 0(3)+2(5) & 0(-1)+2(4) & 0(6)+2(1)\end{bmatrix}$$
$$=\begin{bmatrix}2 & 18 & 11 & 9\\ 0 & 10 & 8 & 2\end{bmatrix}$$

6c. Find the product, if possible.

$$\begin{bmatrix}2 & 3 & -1 & 6\\ 0 & 5 & 4 & 1\end{bmatrix}\begin{bmatrix}1 & 3\\ 0 & 2\end{bmatrix}$$

The number of columns in the first matrix does not equal the number of rows in the second matrix. The product of the matrices is undefined.

Pencil Problem #6

6a. Find AB, given $A=\begin{bmatrix}1 & 3\\ 5 & 3\end{bmatrix}$ and $B=\begin{bmatrix}3 & -2\\ -1 & 6\end{bmatrix}$.

6b. Find the product, if possible.

$$\begin{bmatrix}4 & 2\\ 6 & 1\\ 3 & 5\end{bmatrix}\begin{bmatrix}2 & 3 & 4\\ -1 & -2 & 0\end{bmatrix}$$

6c. Find the product, if possible.

$$\begin{bmatrix}2 & 3 & 4\\ -1 & -2 & 0\end{bmatrix}\begin{bmatrix}4 & 2\\ 6 & 1\\ 3 & 5\end{bmatrix}$$

Objective #7: Model applied situations with matrix operations.

✔ *Solved Problem #7*

7. The triangle with vertices (0, 0), (3, 5), and (4, 2) in a rectangular coordinate system can be represented by the matrix $\begin{bmatrix} 0 & 3 & 4 \\ 0 & 5 & 2 \end{bmatrix}$. Use matrix operations to move the triangle 3 units to the left and 1 unit down. Graph the original triangle and the transformed triangle in the same rectangular coordinate system.

We subtract 3 from each x-coordinate and subtract 1 from each y-coordinate.

$$\begin{bmatrix} 0 & 3 & 4 \\ 0 & 5 & 2 \end{bmatrix} + \begin{bmatrix} -3 & -3 & -3 \\ -1 & -1 & -1 \end{bmatrix}$$

$$= \begin{bmatrix} -3 & 0 & 1 \\ -1 & 4 & 1 \end{bmatrix}$$

The vertices of the translated triangle are (–3, –4), (0, 4), and (1, 1).

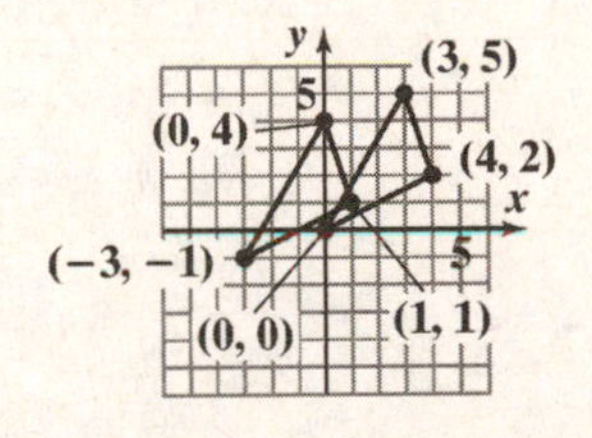

Pencil Problem #7

7. An L-shaped figure has vertices at (0, 0), (3, 0), (3, 1), (1, 1), (1, 5), and (0, 5) in a rectangular coordinate system and can be represented by the matrix $\begin{bmatrix} 0 & 3 & 3 & 1 & 1 & 0 \\ 0 & 0 & 1 & 1 & 5 & 5 \end{bmatrix}$. Use matrix operations to move the figure 2 units to the left and 3 units down.
Graph the original figure and the transformed figure in the same rectangular coordinate system.

Answers for Pencil Problems *(Textbook Exercise references in parentheses)*:

1a. 3×4 *(6.3 #3a)* **1b.** $a_{32} = \frac{1}{2}$; $a_{23} = -6$ *(6.3 #3b)*

2. $x = 4, y = 6, z = 3$ (6.3 #7)

3a. $\begin{bmatrix} 3 & 2 \\ 6 & 2 \\ 5 & 7 \end{bmatrix}$ *(6.3 #11a)* **3b.** $\begin{bmatrix} -1 & -8 \\ 3 & -5 \end{bmatrix}$ *(6.3 #9b)*

4a. $\begin{bmatrix} -8 \\ 16 \\ -4 \end{bmatrix}$ *(6.3 #13c)* **4b.** $\begin{bmatrix} -4 \\ -6 \\ 1 \end{bmatrix}$ *(6.3 #13d)*

5. $X = \begin{bmatrix} -1 & 3 \\ -1 & \frac{9}{2} \\ -1 & -2 \end{bmatrix}$ *(6.3 #19)*

6a. $\begin{bmatrix} 0 & 16 \\ 12 & 8 \end{bmatrix}$ *(6.3 #27a)* **6b.** $\begin{bmatrix} 6 & 8 & 16 \\ 11 & 16 & 24 \\ 1 & -1 & 12 \end{bmatrix}$ *(6.3 #33a)* **6c.** $\begin{bmatrix} 38 & 27 \\ -16 & -4 \end{bmatrix}$ *(6.3 #33b)*

7. $\begin{bmatrix} -2 & 1 & 1 & -1 & -1 & -2 \\ -3 & -3 & -2 & -2 & 2 & 2 \end{bmatrix}$;

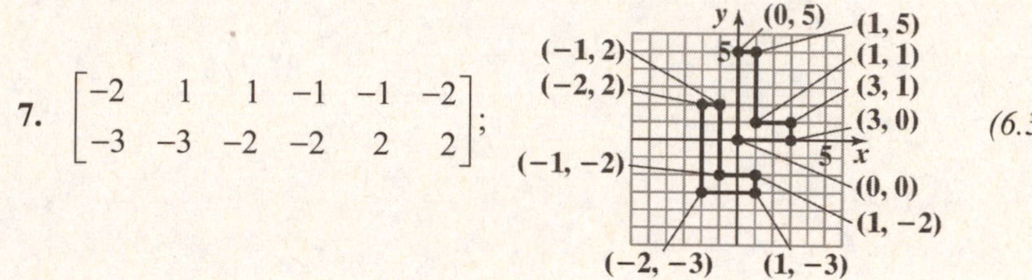

(6.3 #53)

Section 6.4
Multiplicative Inverses of Matrices and Matrix Equations

Think you're too old for sending secret messages?

Did you know that a secure electronic message is usually encrypted in such a way that only the receiver will be able to decrypt it to read it? The technology used is a bit more sophisticated than the cereal-box ciphers that we used to send secret messages as kids.

Matrix multiplication can be used to encrypt a message. To decrypt the message, we need to know the multiplicative inverse of the encryption matrix. Without the original matrix or its inverse, it is extremely difficult to "break the code."

Objective #1: Find the multiplicative inverse of a square matrix.

✔ *Solved Problem #1*

1a. Find the multiplicative inverse of $A = \begin{bmatrix} 3 & -2 \\ -1 & 1 \end{bmatrix}$.

$A = \begin{bmatrix} a & b \\ c & d \end{bmatrix} = \begin{bmatrix} 3 & -2 \\ -1 & 1 \end{bmatrix}$, so $a = 3$, $b = -2$, $c = -1$, and $d = 1$.

$ad - bc = 3(1) - (-2)(-1) = 1 \neq 0$, so the matrix has an inverse.

Using the quick method,

$$A^{-1} = \frac{1}{ad - bc}\begin{bmatrix} d & -b \\ -c & a \end{bmatrix}$$

$$= \frac{1}{3(1) - (-2)(-1)}\begin{bmatrix} 1 & -(-2) \\ -(-1) & 3 \end{bmatrix}$$

$$= \frac{1}{1}\begin{bmatrix} 1 & 2 \\ 1 & 3 \end{bmatrix} = \begin{bmatrix} 1 & 2 \\ 1 & 3 \end{bmatrix}$$

You can verify the result by showing that $AA^{-1} = I_2$ and $A^{-1}A = I_2$, where $I_2 = \begin{bmatrix} 1 & 0 \\ 0 & 1 \end{bmatrix}$ is the 2 × 2 identity matrix.

Pencil Problem #1

1a. Find the multiplicative inverse of $A = \begin{bmatrix} 2 & 3 \\ -1 & 2 \end{bmatrix}$.

1b. Find the multiplicative inverse of $A=\begin{bmatrix}1 & 0 & 2\\ -1 & 2 & 3\\ 1 & -1 & 0\end{bmatrix}$.

Form the augmented matrix $[A|I_3]$ and perform row operations to obtain a matrix of the form $[I_3|B]$.

$$\left[\begin{array}{ccc|ccc}1 & 0 & 2 & 1 & 0 & 0\\ -1 & 2 & 3 & 0 & 1 & 0\\ 1 & -1 & 0 & 0 & 0 & 1\end{array}\right]$$

$$\xrightarrow[-R_1+R_3]{R_1+R_2}\left[\begin{array}{ccc|ccc}1 & 0 & 2 & 1 & 0 & 0\\ 0 & 2 & 5 & 1 & 1 & 0\\ 0 & -1 & -2 & -1 & 0 & 1\end{array}\right]$$

$$\xrightarrow{\frac{1}{2}R_2}\left[\begin{array}{ccc|ccc}1 & 0 & 2 & 1 & 0 & 0\\ 0 & 1 & \frac{5}{2} & \frac{1}{2} & \frac{1}{2} & 0\\ 0 & -1 & -2 & -1 & 0 & 1\end{array}\right]$$

$$\xrightarrow{R_2+R_3}\left[\begin{array}{ccc|ccc}1 & 0 & 2 & 1 & 0 & 0\\ 0 & 1 & \frac{5}{2} & \frac{1}{2} & \frac{1}{2} & 0\\ 0 & 0 & \frac{1}{2} & -\frac{1}{2} & \frac{1}{2} & 1\end{array}\right]$$

$$\xrightarrow{2R_3}\left[\begin{array}{ccc|ccc}1 & 0 & 2 & 1 & 0 & 0\\ 0 & 1 & \frac{5}{2} & \frac{1}{2} & \frac{1}{2} & 0\\ 0 & 0 & 1 & -1 & 1 & 2\end{array}\right]$$

$$\xrightarrow[-\frac{5}{2}R_3+R_2]{-2R_3+R_1}\left[\begin{array}{ccc|ccc}1 & 0 & 0 & 3 & -2 & -4\\ 0 & 1 & 0 & 3 & -2 & -5\\ 0 & 0 & 1 & -1 & 1 & 2\end{array}\right]$$

The inverse matrix is

$$A^{-1}=\begin{bmatrix}3 & -2 & -4\\ 3 & -2 & -5\\ -1 & 1 & 2\end{bmatrix}.$$

You can verify the result by showing that $AA^{-1}=I_3$ and $A^{-1}A=I_3$.

1b. Find the multiplicative inverse of

$$A=\begin{bmatrix}1 & 2 & -1\\ -2 & 0 & 1\\ 1 & -1 & 0\end{bmatrix}.$$

Objective #2: Use inverses to solve matrix equations.

✔ *Solved Problem #2*

2. Solve the system by using A^{-1}, the inverse of the coefficient matrix, where $A^{-1} = \begin{bmatrix} 3 & -2 & -4 \\ 3 & -2 & -5 \\ -1 & 1 & 2 \end{bmatrix}$.

$$\begin{cases} x + 2z = 6 \\ -x + 2y + 3z = -5 \\ x - y = 6 \end{cases}$$

The system can be written as

$$\begin{bmatrix} 1 & 0 & 2 \\ -1 & 2 & 3 \\ 1 & -1 & 0 \end{bmatrix}\begin{bmatrix} x \\ y \\ z \end{bmatrix} = \begin{bmatrix} 6 \\ -5 \\ 6 \end{bmatrix},$$

which is of the form $AX = B$. The solution is $X = A^{-1}B$.

$$X = A^{-1}B = \begin{bmatrix} 3 & -2 & -4 \\ 3 & -2 & -5 \\ -1 & 1 & 2 \end{bmatrix}\begin{bmatrix} 6 \\ -5 \\ 6 \end{bmatrix}$$

$$= \begin{bmatrix} 3(6) - 2(-5) - 4(6) \\ 3(6) - 2(-5) - 5(6) \\ -1(6) + 1(-5) + 2(6) \end{bmatrix}$$

$$= \begin{bmatrix} 4 \\ -2 \\ 1 \end{bmatrix}$$

So, $x = 4$, $y = -2$, and $z = 1$.
The solution set is $\{(4, -2, 1)\}$.

Pencil Problem #2

2. Solve the system by using A^{-1}, the inverse of the coefficient matrix, where

$$A^{-1} = \begin{bmatrix} 3 & 3 & -1 \\ -2 & -2 & 1 \\ -4 & -5 & 2 \end{bmatrix}.$$

$$\begin{cases} x - y + z = 8 \\ 2y - z = -7 \\ 2x + 3y = 1 \end{cases}$$

Objective #3: Encode and decode messages.

✔ *Solved Problem #3*

3a. Use the coding matrix $\begin{bmatrix} -2 & -3 \\ 3 & 4 \end{bmatrix}$ to encode the word BASE.

The numerical equivalent of the word BASE is 2, 1, 19, 5.

The matrix for the word BASE is $\begin{bmatrix} 2 & 19 \\ 1 & 5 \end{bmatrix}$.

Multiply using the encoding matrix on the left.

$$\begin{bmatrix} -2 & -3 \\ 3 & 4 \end{bmatrix}\begin{bmatrix} 2 & 19 \\ 1 & 5 \end{bmatrix} = \begin{bmatrix} -2(2)-3(1) & -2(19)-3(5) \\ 3(2)+4(1) & 3(19)+4(5) \end{bmatrix}$$

$$= \begin{bmatrix} -7 & -53 \\ 10 & 77 \end{bmatrix}$$

The encoded message is –7, 10, –53, 77.

Pencil Problem #3

3a. Use the coding matrix $\begin{bmatrix} 4 & -1 \\ -3 & 1 \end{bmatrix}$ to encode the word HELP.

3b. Decode the word encoded in Solved Problem 3a.
Find the inverse of the coding matrix.

$$A^{-1} = \frac{1}{-2(4)-(-3)(3)}\begin{bmatrix} 4 & -(-3) \\ -3 & -2 \end{bmatrix}$$

$$= \frac{1}{1}\begin{bmatrix} 4 & 3 \\ -3 & -2 \end{bmatrix} = \begin{bmatrix} 4 & 3 \\ -3 & -2 \end{bmatrix}$$

Multiply A^{-1} and the coded matrix from Solved Problem 3a.

$$\begin{bmatrix} 4 & 3 \\ -3 & -2 \end{bmatrix}\begin{bmatrix} -7 & -53 \\ 10 & 77 \end{bmatrix}$$

$$= \begin{bmatrix} 4(-7)+3(10) & 4(-53)+3(77) \\ -3(-7)-2(10) & -3(-53)-2(77) \end{bmatrix}$$

$$= \begin{bmatrix} 2 & 19 \\ 1 & 5 \end{bmatrix}$$

The decoded message is 2, 1, 19, 5, or BASE.

3b. Decode the word encoded in Pencil Problem 3a.

Answers for Pencil Problems *(Textbook Exercise references in parentheses)*:

1a. $A^{-1} = \frac{1}{7}\begin{bmatrix} 2 & -3 \\ 1 & 2 \end{bmatrix} = \begin{bmatrix} \frac{2}{7} & -\frac{3}{7} \\ \frac{1}{7} & \frac{2}{7} \end{bmatrix}$ *(6.4 #13)* **1b.** $A^{-1} = \begin{bmatrix} 1 & 1 & 2 \\ 1 & 1 & 1 \\ 2 & 3 & 4 \end{bmatrix}$ *(6.4 #21)*

2. {(2, –1, 5)} *(6.4 #39)* **3a.** 27, –19, 32, –20 *(6.4 #51)* **3b.** 8, 5, 12, 16 or HELP *(6.4 #51)*

Section 6.5
Determinants and Cramer's Rule

Look……Closer!!!

Do you see the difference between these two mathematical expressions?

$$\begin{bmatrix} 1 & 2 \\ 0 & 1 \end{bmatrix} \qquad \begin{vmatrix} 1 & 2 \\ 0 & 1 \end{vmatrix}$$

If you look carefully, you will notice that the expression on the left is surrounded by brackets, [ ], and is therefore a **matrix**. The expression on the right is surrounded by bars, | |, and represents a **determinant**.

But be careful! This section will discuss *both* determinants *and* matrices.

Objective #1: Evaluate a second-order determinant.

✔ Solved Problem #1

1. Evaluate the determinant of the matrix: $\begin{bmatrix} 10 & 9 \\ 6 & 5 \end{bmatrix}$

The determinant of the matrix $\begin{bmatrix} 10 & 9 \\ 6 & 5 \end{bmatrix}$ is $\begin{vmatrix} 10 & 9 \\ 6 & 5 \end{vmatrix}$.

$$\begin{vmatrix} 10 & 9 \\ 6 & 5 \end{vmatrix} = 10(5) - 6(9) = 50 - 54 = -4$$

Pencil Problem #1

1. Evaluate the determinant: $\begin{vmatrix} -4 & 1 \\ 5 & 6 \end{vmatrix}$

Objective #2: Solve a system of linear equations in two variables using Cramer's rule.

✔ Solved Problem #2

2. Use Cramer's rule to solve the system:
$$\begin{cases} 5x + 4y = 12 \\ 3x - 6y = 24 \end{cases}$$

$$D = \begin{vmatrix} 5 & 4 \\ 3 & -6 \end{vmatrix} = 5(-6) - 3(4) = -30 - 12 = -42$$

$$D_x = \begin{vmatrix} 12 & 4 \\ 24 & -6 \end{vmatrix} = 12(-6) - 24(4) = -72 - 96 = -168$$

$$D_y = \begin{vmatrix} 5 & 12 \\ 3 & 24 \end{vmatrix} = 5(24) - 3(12) = 120 - 36 = 84$$

$$x = \frac{D_x}{D} = \frac{-168}{-42} = 4 \quad y = \frac{D_y}{D} = \frac{84}{-42} = -2$$

The solution set is $\{(4, -2)\}$.

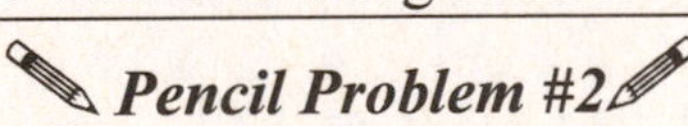

Pencil Problem #2

2. Use Cramer's rule to solve the system:
$$\begin{cases} 12x + 3y = 15 \\ 2x - 3y = 13 \end{cases}$$

Objective #3: Evaluate a third-order determinant.

Solved Problem #3

3. Evaluate the determinant of the matrix:

$$\begin{bmatrix} 2 & 1 & 7 \\ -5 & 6 & 0 \\ -4 & 3 & 1 \end{bmatrix}$$

$$\begin{vmatrix} 2 & 1 & 7 \\ -5 & 6 & 0 \\ -4 & 3 & 1 \end{vmatrix}$$

$$= 2\begin{vmatrix} 6 & 0 \\ 3 & 1 \end{vmatrix} - (-5)\begin{vmatrix} 1 & 7 \\ 3 & 1 \end{vmatrix} - 4\begin{vmatrix} 1 & 7 \\ 6 & 0 \end{vmatrix}$$

$$= 2(6(1)-3(0))+5(1(1)-3(7))-4(1(0)-6(7))$$

$$= 2(6)+5(-20)-4(-42)$$

$$= 12-100+168$$

$$= 80$$

Pencil Problem #3

3. Evaluate the determinant:

$$\begin{vmatrix} 3 & 0 & 0 \\ 2 & 1 & -5 \\ 2 & 5 & -1 \end{vmatrix}$$

Objective #4: Solve a system of linear equations in three variables using Cramer's rule.

Solved Problem #4

4. Use Cramer's rule to solve the system:

$$\begin{cases} 3x-2y+z=16 \\ 2x+3y-z=-9 \\ x+4y+3z=2 \end{cases}$$

First, find D, D_x, D_y, and D_z.

$$D = \begin{vmatrix} 3 & -2 & 1 \\ 2 & 3 & -1 \\ 1 & 4 & 3 \end{vmatrix}$$

$$= 3\begin{vmatrix} 3 & -1 \\ 4 & 3 \end{vmatrix} - 2\begin{vmatrix} -2 & 1 \\ 4 & 3 \end{vmatrix} + 1\begin{vmatrix} -2 & 1 \\ 3 & -1 \end{vmatrix}$$

$$= 58$$

(Continued on next page)

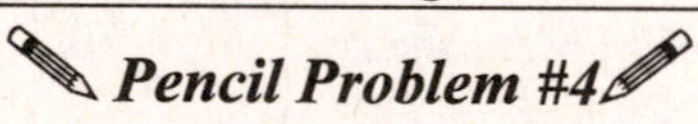

Pencil Problem #4

4. Use Cramer's rule to solve the system:

$$\begin{cases} x+y+z=0 \\ 2x-y+z=-1 \\ -x+3y-z=-8 \end{cases}$$

$$D_x = \begin{vmatrix} 16 & -2 & 1 \\ -9 & 3 & -1 \\ 2 & 4 & 3 \end{vmatrix}$$

$$= 16\begin{vmatrix} 3 & -1 \\ 4 & 3 \end{vmatrix} - (-9)\begin{vmatrix} -2 & 1 \\ 4 & 3 \end{vmatrix} + 2\begin{vmatrix} -2 & 1 \\ 3 & -1 \end{vmatrix}$$

$$= 116$$

$$D_y = \begin{vmatrix} 3 & 16 & 1 \\ 2 & -9 & -1 \\ 1 & 2 & 3 \end{vmatrix}$$

$$= 3\begin{vmatrix} -9 & -1 \\ 2 & 3 \end{vmatrix} - 2\begin{vmatrix} 16 & 1 \\ 2 & 3 \end{vmatrix} + 1\begin{vmatrix} 16 & 1 \\ -9 & -1 \end{vmatrix}$$

$$= -174$$

$$D_z = \begin{vmatrix} 3 & -2 & 16 \\ 2 & 3 & -9 \\ 1 & 4 & 2 \end{vmatrix}$$

$$= 3\begin{vmatrix} 3 & -9 \\ 4 & 2 \end{vmatrix} - 2\begin{vmatrix} -2 & 16 \\ 4 & 2 \end{vmatrix} + 1\begin{vmatrix} -2 & 16 \\ 3 & -9 \end{vmatrix}$$

$$= 232$$

Next, use D, D_x, D_y, and D_z to find x, y, and z.

$D = 58$, $D_x = 116$, $D_y = -174$, and $D_z = 232$.

$$x = \frac{D_x}{D} = \frac{116}{58} = 2$$

$$y = \frac{D_y}{D} = \frac{-174}{58} = -3$$

$$z = \frac{D_z}{D} = \frac{232}{58} = 4$$

The solution set is $\{(2,-3,4)\}$.

Objective #5: Evaluate higher-order determinants.

✔ Solved Problem #5

5. Evaluate the determinant: $\begin{vmatrix} 0 & 4 & 0 & -3 \\ -1 & 1 & 5 & 2 \\ 1 & -2 & 0 & 6 \\ 3 & 0 & 0 & 1 \end{vmatrix}$.

With three 0s in the third column, we expand along the third column.

$$\begin{vmatrix} 0 & 4 & 0 & -3 \\ -1 & 1 & 5 & 2 \\ 1 & -2 & 0 & 6 \\ 3 & 0 & 0 & 1 \end{vmatrix}$$

$$= (-1)^{2+3}(5)\begin{vmatrix} 0 & 4 & -3 \\ 1 & -2 & 6 \\ 3 & 0 & 1 \end{vmatrix}$$

$$= -5\left((-1)^{2+1}(1)\begin{vmatrix} 4 & -3 \\ 0 & 1 \end{vmatrix} + (-1)^{3+1}(3)\begin{vmatrix} 4 & -3 \\ -2 & 6 \end{vmatrix}\right)$$

$$= -5\left(-1(4(1)-(-3)(0))+3(4(6)-(-3)(-2))\right)$$

$$= -5(-1(4-0)+3(24-6))$$

$$= -5(-4+54)$$

$$= -5(50)$$

$$= -250$$

Pencil Problem #5

5. Evaluate the determinant: $\begin{vmatrix} 4 & 2 & 8 & -7 \\ -2 & 0 & 4 & 1 \\ 5 & 0 & 0 & 5 \\ 4 & 0 & 0 & -1 \end{vmatrix}$.

Answers for Pencil Problems *(Textbook Exercise references in parentheses)*:

1. -29 *(6.5 #3)*

2. $\{(2,-3)\}$ *(6.5 #13)*

3. 72 *(6.5 #23)*

4. $\{(-5,-2,7)\}$ *(6.5 #29)*

5. -200 *(6.5 #37)*

Section 7.1
The Ellipse

Do You Trust Politicians?

The U.S. Capitol Building is beautiful both inside, and out. But did you know that part of its architecture includes an elliptical ceiling in Sanctuary Hall?

John Quincy Adams, while a member of the house of Representatives, discovered that he could use the reflective properties of the room, which we will study in this section, to eavesdrop on the conversations of other House members.

Objective #1: Graph ellipses centered at the origin.

✔ Solved Problem #1

1. Graph and locate the foci: $16x^2+9y^2=144$

First, write the equation in standard form.

$$16x^2+9y^2=144$$

$$\frac{16x^2}{144}+\frac{9y^2}{144}=\frac{144}{144}$$

$$\frac{x^2}{9}+\frac{y^2}{16}=1$$

Because the denominator of the y^2 – term is greater than the denominator of the x^2 – term, the major axis is vertical.

Since $a^2=16$, $a=4$ and the vertices are $(0,-4)$ and $(0,4)$.

Since $b^2=9$, $b=3$ and endpoints of the minor axis are $(-3,0)$ and $(3,0)$.

$c^2=a^2-b^2=16-9=7$, $c=\sqrt{7}$ and the foci are $(0,-\sqrt{7})$ and $(0,\sqrt{7})$.

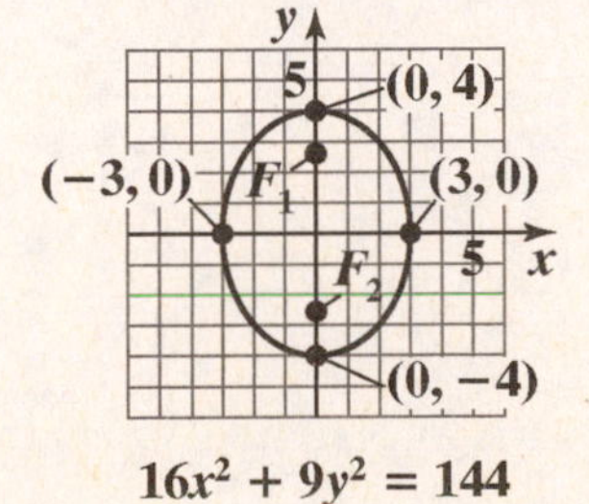

Pencil Problem #1

1. Graph and locate the foci: $\frac{x^2}{16}+\frac{y^2}{4}=1$

Objective #2: Write equations of ellipses in standard form.

✔ *Solved Problem #2*

2. Find the standard form of the equation of an ellipse with foci at (–2, 0) and (2, 0) and vertices (–3, 0) and (3, 0).

Because the foci are located on the x-axis, the major axis is horizontal with the center midway between them at (0, 0). The form of the equation is $\frac{x^2}{a^2}+\frac{y^2}{b^2}=1$. We need to determine values for a^2 and b^2. The distance from the center to either vertex is 3, so $a = 3$ and $a^2 = 9$. The distance from the center to either focus is 2, so $c = 2$.

$$b^2 = a^2 - c^2 = 3^2 - 2^2 = 5$$

The equation is $\frac{x^2}{9}+\frac{y^2}{5}=1$.

Pencil Problem #2

2. Find the standard form of the equation of an ellipse with foci at (0, –4) and (0, 4) and vertices (0, –7) and (0, 7).

Objective #3: Graph ellipses not centered at the origin.

✔ *Solved Problem #3*

3. Graph: $\frac{(x+1)^2}{9}+\frac{(y-2)^2}{4}=1$. Where are the foci located?

$$\frac{(x+1)^2}{9}+\frac{(y-2)^2}{4}=1$$

The center of the ellipse is $(-1,2)$.

Because the denominator of the x^2 – term is greater than the denominator of the y^2 – term, the major axis is horizontal.

Since $a^2 = 9$, $a = 3$ and the vertices lie 3 units to the right and left of the center.

Since $b^2 = 4$, $b = 2$ and endpoints of the minor axis lie 2 units above and below the center.

Since $c^2 = a^2 - b^2 = 9 - 4 = 5$, $c = \sqrt{5}$ and the foci are located $\sqrt{5}$ units to the right and left of center.

The following chart summarizes these key points.

(Continued on next page)

Pencil Problem #3

3. Graph: $\frac{(x-4)^2}{9}+\frac{(y+2)^2}{25}=1$. Where are the foci located?

Center	Vertices	Endpoints Minor Axis	Foci
$(-1,2)$	$(-1-3,2)$ $=(-4,2)$	$(-1,2-2)$ $=(-1,0)$	$(-1-\sqrt{5},2)$
	$(-1+3,2)$ $=(2,2)$	$(-1,2+2)$ $=(-1,4)$	$(-1+\sqrt{5},2)$

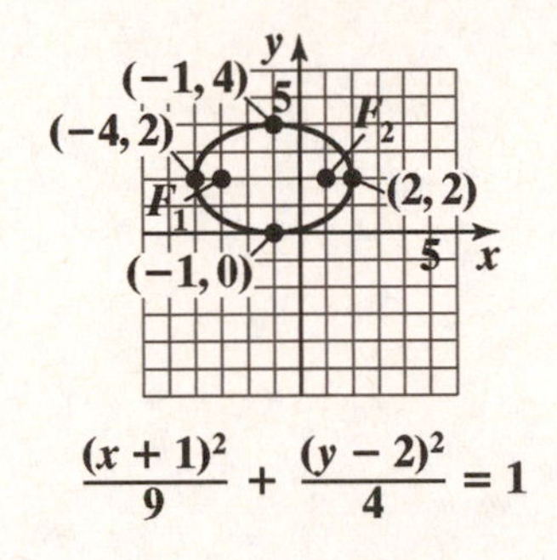

$$\frac{(x+1)^2}{9}+\frac{(y-2)^2}{4}=1$$

Objective #4: Solve applied problems involving ellipses.

✔ Solved Problem #4

4. A semielliptical archway over a one-way road has a height of 10 feet and a width of 40 feet. Your truck has a width of 12 feet and a height of 9 feet. Will your truck clear the opening of the archway?

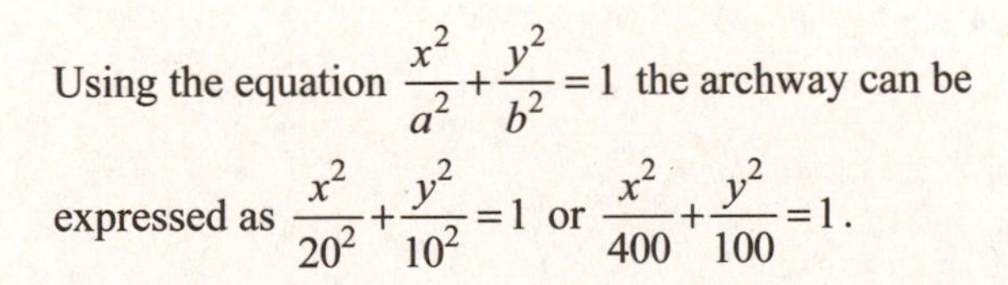

Using the equation $\frac{x^2}{a^2}+\frac{y^2}{b^2}=1$ the archway can be expressed as $\frac{x^2}{20^2}+\frac{y^2}{10^2}=1$ or $\frac{x^2}{400}+\frac{y^2}{100}=1$.

Since the truck is 12 feet wide, we need to determine the height of the archway at $\frac{12}{2}=6$ feet from the center.

Substitute 6 for x to find the height y.

$$\frac{x^2}{400}+\frac{y^2}{100}=1$$

$$\frac{6^2}{400}+\frac{y^2}{100}=1$$

(Continued on next page)

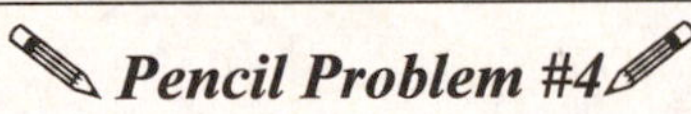

Pencil Problem #4

4. Will a truck that is 8 feet wide carrying a load that reaches 7 feet above the ground clear the semielliptical arch on the one-way road that passes under a bridge that has a height of 10 feet and a width of 30 feet?

Solve for y.

$$\frac{6^2}{400}+\frac{y^2}{100}=1$$

$$\frac{36}{400}+\frac{y^2}{100}=1$$

$$400\left(\frac{36}{400}+\frac{y^2}{100}\right)=400(1)$$

$$36+4y^2=400$$

$$4y^2=364$$

$$y^2=91$$

$$y=\sqrt{91}\approx 9.54$$

The height of the archway 6 feet from the center is approximately 9.54 feet.

Since the truck is 9 feet high, the truck will clear the archway.

Answers for Pencil Problems *(Textbook Exercise references in parentheses)*:

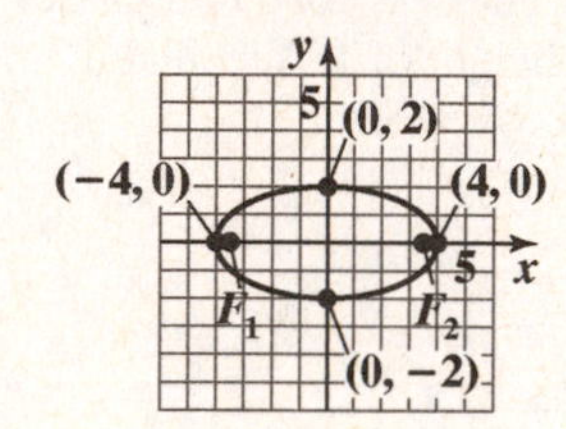

1. $\frac{x^2}{16}+\frac{y^2}{4}=1$ foci at $(-2\sqrt{3},0)$ and $(2\sqrt{3},0)$ *(7.1 #1)*

2. $\frac{x^2}{33}+\frac{y^2}{49}=1$ *(7.1 #27)*

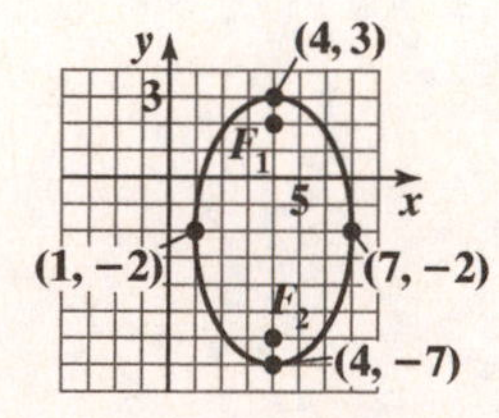

3. $\frac{(x-4)^2}{9}+\frac{(y+2)^2}{25}=1$ foci at (4, 2) and (4, −6) *(7.1 #41)*

4. Yes; the height of the archway 4 feet from the center is approximately 9.64 feet. *(7.1 #65)*

Section 7.2
The Hyperbola

Sonic Boom!

When a jet flies at a speed greater than the speed of sound, the shock wave that is created is heard as a sonic boom.

The wave has the shape of a cone.
The shape formed as the cone hits the ground is one branch of a hyperbola, the topic of this section of the textbook.

Objective #1: Locate a hyperbola's vertices and foci.

✔ *Solved Problem #1*

1a. Find the vertices and locate the foci for the hyperbola with the given equation: $\frac{x^2}{25}-\frac{y^2}{16}=1.$

The x^2 – term is positive.
Therefore, the transverse axis lies along the x–axis.

Since $a^2=25$ and $a=5,$ the vertices are $(-5,0)$ and $(5,0)$.

Since $c^2=a^2+b^2=25+16=41,$ $c=\sqrt{41}$ and the foci are $(-\sqrt{41},0)$ and $(\sqrt{41},0)$.

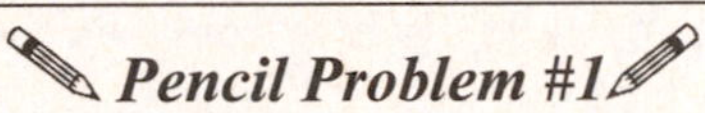

1a. Find the vertices and locate the foci for the hyperbola with the given equation: $\frac{x^2}{4}-\frac{y^2}{1}=1.$

1b. Find the vertices and locate the foci for the hyperbola with the given equation: $\frac{y^2}{25}-\frac{x^2}{16}=1.$

The y^2 – term is positive.
Therefore, the transverse axis lies along the y–axis.

Since $a^2=25$ and $a=5,$ the vertices are $(0,-5)$ and $(0,5)$.

Since $c^2=a^2+b^2=25+16=41,$ $c=\sqrt{41}$ and the foci are $(0,-\sqrt{41})$ and $(0,\sqrt{41})$.

1b. Find the vertices and locate the foci for the hyperbola with the given equation: $\frac{y^2}{4}-\frac{x^2}{1}=1.$

Objective #2: Write equations of hyperbolas in standard form.

✔ *Solved Problem #2*

2. Find the standard form of the equation of a hyperbola with foci at (0, –5) and (0, 5) and vertices (0, –3) and (0, 3).

Because the foci are located on the y-axis, the transverse axis lies on the y-axis with the center midway between the foci at (0, 0). The form of the equation is $\frac{y^2}{a^2}-\frac{x^2}{b^2}=1$. We need to determine values for a^2 and b^2.

The distance from the center to either vertex is 3, so $a=3$ and $a^2=9$. The distance from the center to either focus is 5, so $c=5$.

$b^2=c^2-a^2=5^2-3^2=16$

The equation is $\frac{y^2}{9}-\frac{x^2}{16}=1$.

Pencil Problem #2

2. Find the standard form of the equation of a hyperbola with foci at (–4, 0) and (4, 0) and vertices (–3, 0) and (3, 0).

Objective #3: Graph hyperbolas centered at the origin.

✔ *Solved Problem #3*

3a. Graph and locate the foci: $\frac{x^2}{36}-\frac{y^2}{9}=1$. What are the equations of the asymptotes?

$$\frac{x^2}{36}-\frac{y^2}{9}=1$$

Since the x^2 –term is positive, the transverse axis lies along the x–axis.

Since $a^2=36$ and $a=6$, the vertices are $(-6,0)$ and $(6,0)$.

Construct a rectangle using –6 and 6 on the x–axis, and –3 and 3 on the y–axis.

Draw extended diagonals to obtain the asymptotes. The equations of the asymptotes are $y=\pm\frac{3}{6}x=\pm\frac{1}{2}x$.

Draw the two branches of the hyperbola by starting at each vertex and approaching the asymptotes.

(Continued on next page)

Pencil Problem #3

3a. Graph and locate the foci: $\frac{x^2}{9}-\frac{y^2}{25}=1$. What are the equations of the asymptotes?

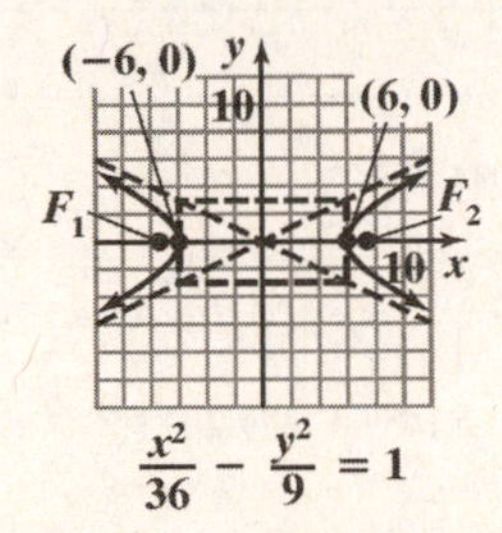

$\frac{x^2}{36} - \frac{y^2}{9} = 1$

Since $c^2 = a^2 + b^2 = 36 + 9 = 45$, $c = \sqrt{45} = 3\sqrt{5}$ and the foci are located at $(-3\sqrt{5}, 0)$ and $(3\sqrt{5}, 0)$.

3b. Graph and locate the foci: $y^2 - 4x^2 = 4$. What are the equations of the asymptotes?

First write the equation in standard form.

$$y^2 - 4x^2 = 4$$

$$\frac{y^2}{4} - \frac{4x^2}{4} = \frac{4}{4}$$

$$\frac{y^2}{4} - \frac{x^2}{1} = 1$$

The equation is in the form $\frac{y^2}{a^2} - \frac{x^2}{b^2} = 1$ with

$a^2 = 4$ and $b^2 = 1$.

The transverse axis lies on the y-axis and the vertices are $(0,-2)$ and $(0,2)$.

Because $a^2 = 4$ and $b^2 = 1$, $a = 2$ and $b = 1$.

Construct a rectangle using -2 and 2 on the y–axis, and -1 and 1 on the x–axis.

Draw extended diagonals to obtain the asymptotes.

The equations of the asymptotes are $y = \pm\frac{2}{1}x = \pm 2x$.

Draw the two branches of the hyperbola by starting at each vertex and approaching the asymptotes.

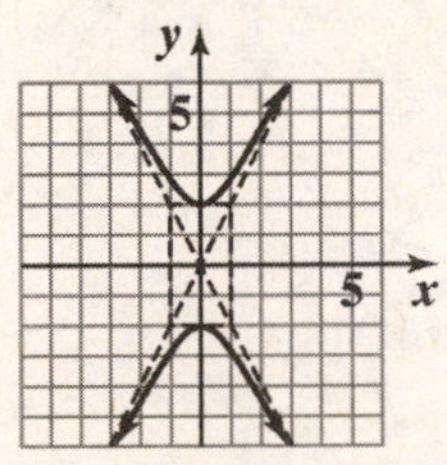

$y^2 - 4x^2 = 4$

Since $c^2 = a^2 + b^2 = 4 + 1 = 5$, $c = \sqrt{5}$ and the foci are located at $(0, -\sqrt{5})$ and $(0, \sqrt{5})$.

3b. Graph and locate the foci: $9y^2 - 25x^2 = 225$. What are the equations of the asymptotes?

Objective #4: Graph hyperbolas not centered at the origin.

✔ *Solved Problem #4*

4. Graph: $\frac{(x-3)^2}{4}-\frac{(y-1)^2}{1}=1$. Where are the foci located? What are the equations of the asymptotes?

Because the term involving x^2 has the positive coefficient, the transverse axis is horizontal. Based on the standard form $\frac{(x-h)^2}{a^2}-\frac{(y-k)^2}{b^2}=1$, we see that $h=3$ and $k=1$ so the center is (3, 1). We also see that $a^2=4$ and $b^2=1$, so $a=2$ and $b=1$.

Since $a=2$, the vertices are 2 units to the left and right of the center at (3 − 2, 1), or (1, 1), and (3 + 2, 1), or (5, 1). Draw a rectangle using the vertices, (1, 1) and (5, 1) and the points $b=1$ unit above and below the center. Draw the extended diagonals to obtain the asymptotes.
The asymptotes of the unshifted hyperbola are $y=\pm\frac{b}{a}x=\pm\frac{1}{2}x$. Thus, the asymptotes of the shifted hyperbola are $y-1=\pm\frac{1}{2}(x-3)$.
Draw the two branches of each hyperbola by starting at each vertex and approaching the asymptotes.

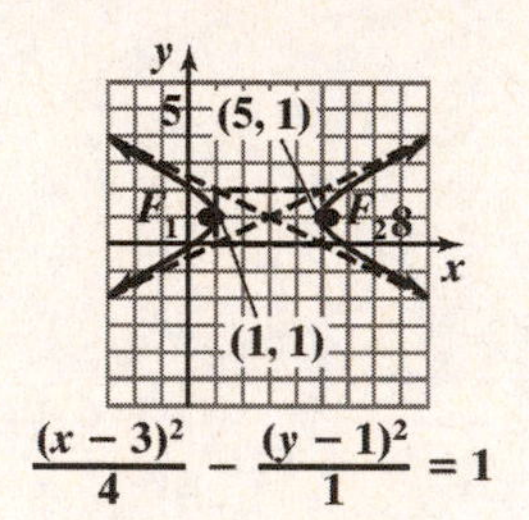

$\frac{(x-3)^2}{4}-\frac{(y-1)^2}{1}=1$

Since $c^2=a^2+b^2=4+1=5$, $c=\sqrt{5}$ and the foci are located $\sqrt{5}$ units to the left and right of center at $(3-\sqrt{5},\ 1)$ and $(3+\sqrt{5},\ 1)$.

Pencil Problem #4

4. Graph: $\frac{(y+2)^2}{4}-\frac{(x-1)^2}{16}=1$. Where are the foci located? What are the equations of the asymptotes?

Objective #5: Solve applied problems involving hyperbolas.

✔ *Solved Problem #5*

5. An explosion is recorded by two microphones that are 2 miles apart. Microphone M_1 received the sound 3 seconds before microphone M_2. Assuming sound travels at 1100 feet per second, determine the possible locations of the explosion relative to the location of the microphones.

Because 1 mile = 5280 feet, place microphone M_1 at (5280, 0) in a coordinate system. Since the microphones are two miles apart, place M_2 at (–5280, 0). Assume that the explosion is at point $P(x, y)$ in the coordinate system. The set of all possible points for the explosion is a hyperbola with the microphones at the foci.

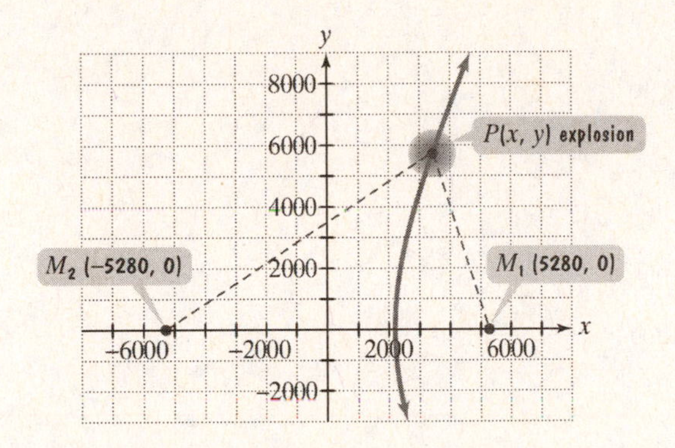

Since M_1 received the sound 3 seconds before microphone M_2 and sound travels at 1100 feet per second, the difference between the distances from P to M_1 and from P to M_2 is 3300 feet. Thus, $2a = 3300$ and $a = 1650$, so $a^2 = 2{,}722{,}500$.

The distance from the center, (0, 0), to either focus is 5280, so $c = 5280$.

$$b^2 = c^2 - a^2 = 5280^2 - 1650^2 = 25{,}155{,}900$$

The equation of the hyperbola is

$\dfrac{x^2}{2{,}722{,}500} - \dfrac{y^2}{25{,}155{,}900} = 1$. The explosion occurred somewhere on the right branch of this hyperbola (the branch closer to M_1).

Pencil Problem #5

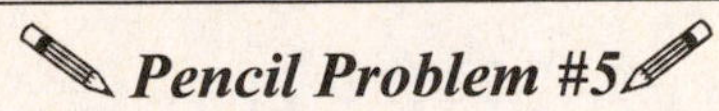

5. An explosion is recorded by two microphones that are 1 mile apart. Microphone M_1 received the sound 2 seconds before microphone M_2. Assuming sound travels at 1100 feet per second, determine the possible locations of the explosion relative to the location of the microphones.

Answers for Pencil Problems *(Textbook Exercise references in parentheses)*:

1a. Vertices: $(-2,0)$ and $(2,0)$; foci: $(-\sqrt{5},0)$ and $(\sqrt{5},0)$ *(7.2 #1)*

1b. Vertices: $(0,-2)$ and $(0,2)$; foci: $(0,-\sqrt{5})$ and $(0,\sqrt{5})$ *(7.2 #3)*

2. $\frac{x^2}{9}-\frac{x^2}{7}=1$ *(7.2 #7)*

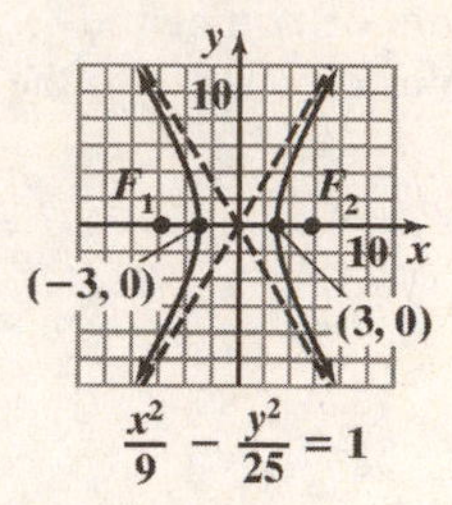

3a. Asymptotes: $y=\pm\frac{5}{3}x$; foci: $(-\sqrt{34},0)$ and $(\sqrt{34},0)$ *(7.2 #13)*

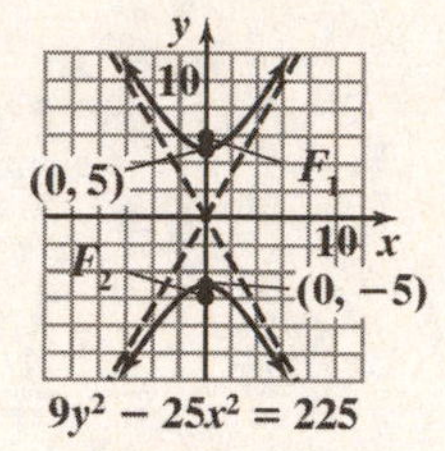

3b. Asymptotes: $y=\pm\frac{5}{3}x$; foci: $(0,-\sqrt{34})$ and $(0,\sqrt{34})$ *(7.2 #23)*

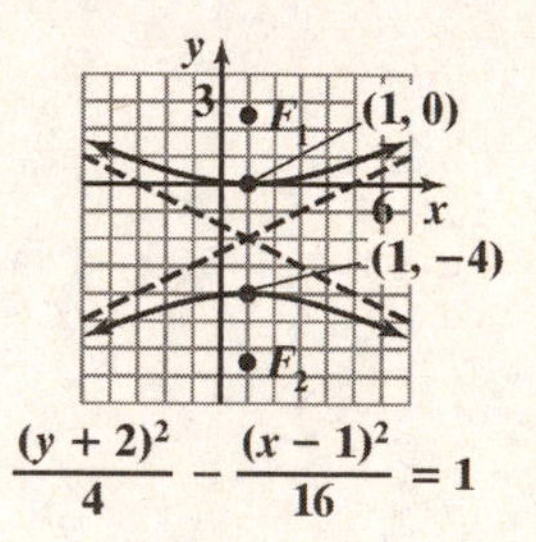

4. Asymptotes: $y+2=\pm\frac{1}{2}(x-1)$; foci: $(1,-2-2\sqrt{5})$ and $(1,-2+2\sqrt{5})$ *(7.2 #37)*

5. If M_1 is located 2640 feet to the right of the origin on the x-axis, the explosion is located on the right branch of the hyperbola given by the equation $\frac{x^2}{1{,}210{,}000}-\frac{x^2}{5{,}759{,}600}=1$. *(7.2 #61)*

Section 7.3
The Parabola

How Good Is Your Reception?

In this section we study parabolas and their properties.
A satellite dish is in the shape of a parabolic surface.
Signals coming from a satellite strike the surface of the dish and are reflected to the focus, where the receiver is located.

The applications in the Exercise Set include concepts from each of the conic sections that we have studied.

Objective #1: Graph parabolas with vertices at the origin.

✔ *Solved Problem #1*

1a. Find the focus and directrix of the parabola given by $y^2 = 8x$. Then graph the parabola.

The equation $y^2 = 8x$ is in the standard form $y^2 = 4px$, so $4p = 8$ and $p = 2$. Because p is positive, the parabola opens to the right. The focus is 2 units to the right of the vertex, (0, 0), at $(p, 0)$ or (2, 0). The directrix is 2 units to the left of the vertex: $x = -p$ or $x = -2$.

To graph the parabola, substitute 2 for x in the equation.

$$y^2 = 8 \cdot 2$$
$$y^2 = 16$$
$$y = \pm\sqrt{16} = \pm 4$$

The points (2, 4) and (2, –4) are on the parabola above and below the focus.

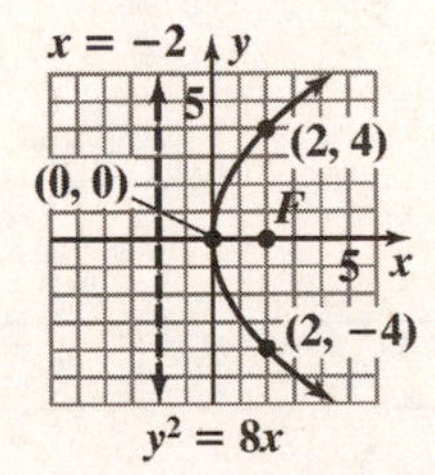

Pencil Problem #1

1a. Find the focus and directrix of the parabola given by $y^2 = 16x$. Then graph the parabola.

1b. Find the focus and directrix of the parabola given by $x^2 = -12y$. Then graph the parabola.

The equation $x^2 = -12y$ is in the standard form $x^2 = 4py$, so $4p = -12$ and $p = -3$. Because p is negative, the parabola opens downward. The focus is 3 units below the vertex, (0, 0), at (0, p) or (0, –3). The directrix is 3 units above the vertex: $y = -p$ or $y = 3$.

To graph the parabola, substitute –3 for y in the equation.

$$x^2 = -12(-3)$$
$$x^2 = 36$$
$$x = \pm\sqrt{36} = \pm 6$$

The points (–6, –3) and (6, –3) are on the parabola to the left and right of the focus.

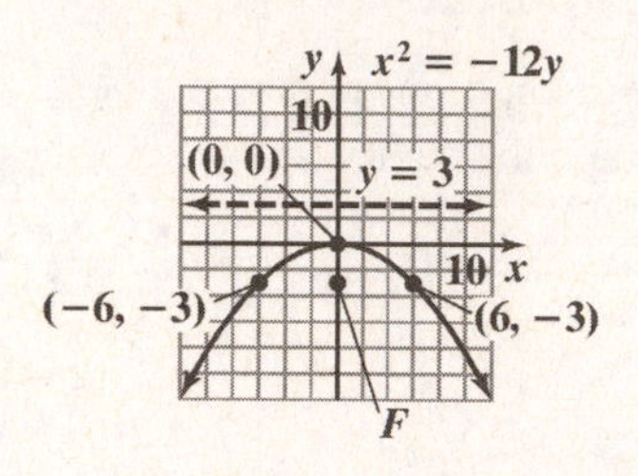

1b. Find the focus and directrix of the parabola given by $x^2 = -16y$. Then graph the parabola.

Objective #2: Write equations of parabolas in standard form.

✔ *Solved Problem #2*

2. Find the standard form of the equation of a parabola with focus (8, 0) and directrix $x = -8$.

The vertex of the parabola is midway between the focus and the directrix at (0, 0). Since the focus is on the x-axis, we use the standard form $y^2 = 4px$.

The focus is 8 units to the right of the vertex, so $p = 8$.

The equation is $y^2 = 4 \cdot 8x$ or $y^2 = 32x$.

Pencil Problem #2

2. Find the standard form of the equation of a parabola with focus (0, 15) and directrix $y = -15$.

Objective #3: Graph parabolas with vertices not at the origin.

✔ *Solved Problem #3*

3a. Find the vertex, focus, and directrix of the parabola given by $(x-2)^2 = 4(y+1)$. Then graph the parabola.

Writing the equation as $(x-2)^2 = 4(y-(-1))$, we see that $h = 2$ and $k = -1$. The vertex is $(h, k) = (2, -1)$.

(Continued on next page)

Pencil Problem #3

3a. Find the vertex, focus, and directrix of the parabola given by $(x+1)^2 = -8(y+1)$. Then graph the parabola.

Because $4p = 4$, $p = 1$. The focus is 1 unit above the vertex at $(h, k + p) = (2, -1 + 1) = (2, 0)$. The directrix is 1 unit below the vertex: $y = k - p = -1 - 1$ or $y = -2$.

The length of the latus rectum is $|4p| = |4 \cdot 1| = |4| = 4$. The latus rectum extends 2 units to the left and right of the focus. The endpoints of the latus rectum are $(2 - 2, 0)$ or $(0, 0)$ and $(2 + 2, 0)$ or $(4, 0)$.

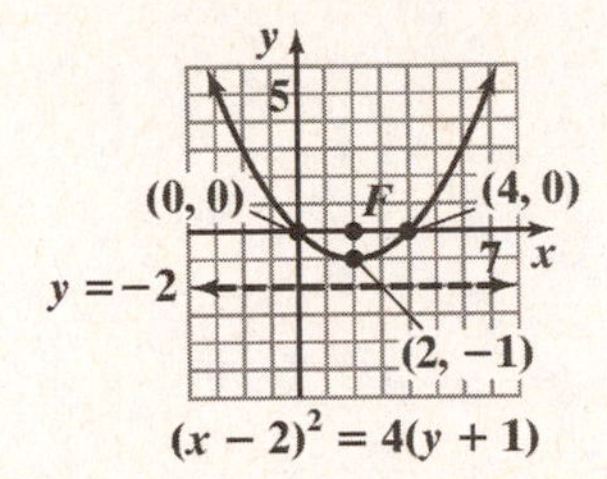

3b. Find the vertex, focus, and directrix of the parabola given by $y^2 + 2y + 4x - 7 = 0$. Then graph the parabola.

Complete the square on y.

$$y^2 + 2y + 4x - 7 = 0$$
$$y^2 + 2y = -4x + 7$$
$$y^2 + 2y + 1 = -4x + 8$$
$$(y + 1)^2 = -4(x - 2)$$

We see that $k = -1$ and $h = 2$, so the vertex is at $(h, k) = (2, -1)$. Since $4p = -4$, $p = -1$. The focus is 1 unit to the left of the vertex at $(h, k) = (2 - 1, -1) = (1, -1)$. The directrix is 1 unit to the right of the vertex: $x = h - p = 2 - (-1)$ or $x = 3$.

The length of the latus rectum is $|4p| = |4(-1)| = |-4| = 4$. The latus rectum extends 2 units above and below the focus. The endpoints of the latus rectum are $(1, -1 + 2)$ or $(1, 1)$ and $(1, -1 - 2)$ or $(1, -3)$.

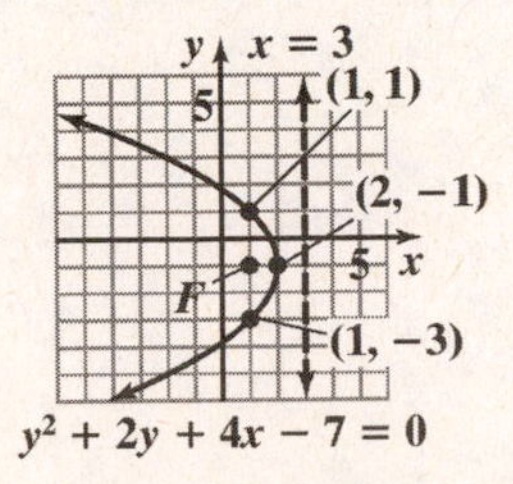

3b. Find the vertex, focus, and directrix of the parabola given by $y^2 - 2y + 12x - 35 = 0$. Then graph the parabola.

Objective #4: Solve applied problems involving parabolas.

✔ Solved Problem #2

4. An engineer is designing a flashlight using a parabolic mirror and a light source. The casting has a diameter of 6 inches and a depth of 4 inches. What is the equation of the parabola used to shape the mirror? At what point should the light source be placed relative to the mirror's vertex?

Position the parabola with its vertex at the origin and opening upward. Then the focus is at $(0, p)$ on the y-axis. Since the casting has a diameter of 6 inches, it extends 3 units to the left and right of the y-axis. Since it is 4 inches deep, the point (3, 4) is on the parabola.

We use the standard form $x^2 = 4py$. Using (3, 4), we have

$$3^2 = 4p \cdot 4$$
$$9 = 16p$$
$$\frac{9}{16} = p.$$

Thus, the equation is $x^2 = 4 \cdot \frac{9}{16}y$ or $x^2 = \frac{9}{4}y$. The light source should be placed at the focus, $(0, p) = (0, \frac{9}{16})$, or $\frac{9}{16}$ inch above the vertex along the axis of symmetry.

Pencil Problem #2

4. The reflector of a flashlight is in the shape of a parabolic surface. The casting has a diameter of 4 inches and a depth of 1 inch. What is the equation of the parabola used to shape the mirror? At what point should the light source be placed relative to the mirror's vertex?

Answers for Pencil Problems *(Textbook Exercise references in parentheses)*:

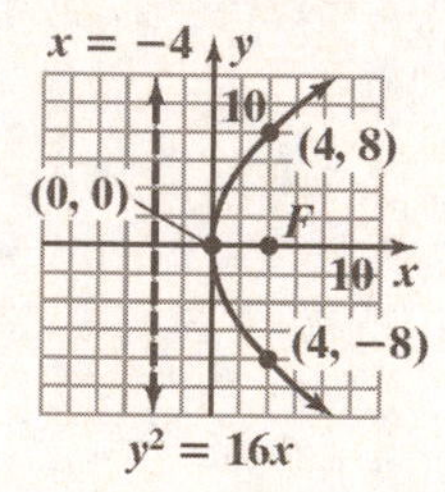

1a. focus: (4, 0); directrix: $x = -4$ *(7.3 #5)*

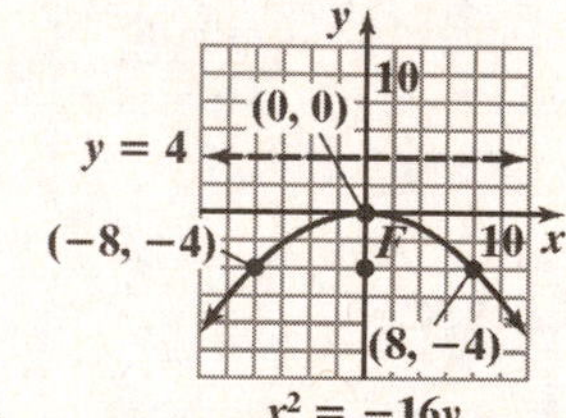

1b. focus: (0, −4); directrix: $y = 4$ *(7.3 #11)*

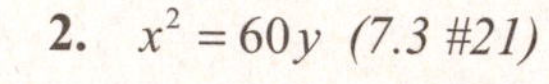

2. $x^2 = 60y$ *(7.3 #21)*

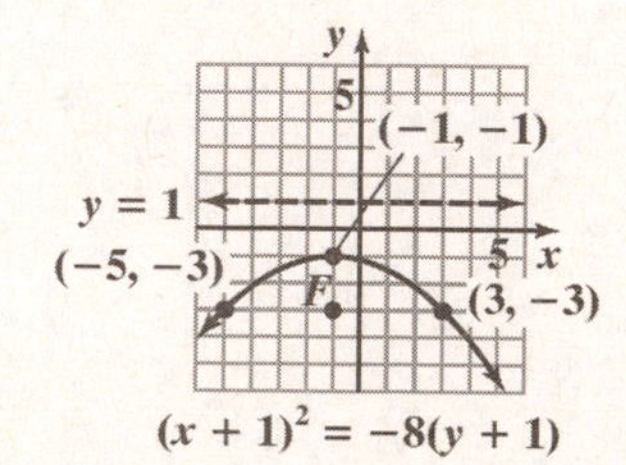

3a. vertex: (−1, −1); focus: (−1, −3); directrix: $y = 1$ *(7.3 #37)*

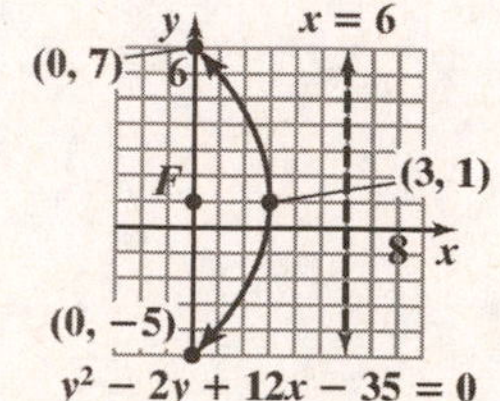

3b. vertex: (3, 1); focus: (0, 1); directrix: $x = 6$ *(7.3 #45)*

4. $x^2 = 4y$; 1 inch above the vertex along the axis of symmetry *(7.3 #61)*

Section 8.1
Sequences and Summation Notation

Bees, Trees, and Piano Keys !

What can those three things possibly have in common?

In this section, we will study sequences. One amazing example is called the Fibonacci sequence, an infinite sequence of numbers investigated by Leonardo of Pisa, also known as Fibonacci, an Italian mathematician of the thirteenth century.

The sequence is generated using simple addition, and yet it shows up in some unexpected, and complex, ways.

As you read the textbook, you will find interesting areas where these concepts apply.

Objective #1: Find particular terms of a sequence from the general term.

✔ *Solved Problem #1*

1a. Write the first four terms of the sequence whose nth term, or general term, is $a_n = 2n+5$.

$$a_n = 2n+5$$
$$a_1 = 2(1)+5 = 7$$
$$a_2 = 2(2)+5 = 9$$
$$a_3 = 2(3)+5 = 11$$
$$a_4 = 2(4)+5 = 13$$

The first four terms are 7, 9, 11, and 13.

1b. Write the first four terms of the sequence whose nth term, or general term, is $a_n = \frac{(-1)^n}{2^n+1}$.

$$a_n = \frac{(-1)^n}{2^n+1}$$
$$a_1 = \frac{(-1)^1}{2^1+1} == \frac{-1}{3} - \frac{1}{3}$$
$$a_2 = \frac{(-1)^2}{2^2+1} = \frac{1}{5}$$
$$a_3 = \frac{(-1)^3}{2^3+1} = \frac{-1}{9} = -\frac{1}{9}$$
$$a_4 = \frac{(-1)^4}{2^4+1} = \frac{1}{17}$$

The first four terms are $-\frac{1}{3}, \frac{1}{5}, -\frac{1}{9},$ and $\frac{1}{17}$.

Pencil Problem #1

1a. Write the first four terms of the sequence whose nth term, or general term, is $a_n = 3n+2$.

1b. Write the first four terms of the sequence whose nth term, or general term, is $a_n = (-1)^n(n+3)$.

Objective #2: Use recursion formulas.

✔ *Solved Problem #2*

2. Find the first four terms of the sequence in which $a_1 = 3$ and $a_n = 2a_{n-1} + 5$ for $n \geq 2$.

$$a_1 = 3$$
$$a_2 = 2a_1 + 5 = 2(3) + 5 = 11$$
$$a_3 = 2a_2 + 5 = 2(11) + 5 = 27$$
$$a_4 = 2a_3 + 5 = 2(27) + 5 = 59$$

The first four terms are 3, 11, 27, and 59.

Pencil Problem #2

2. Find the first four terms of the sequence in which $a_1 = 4$ and $a_n = 2a_{n-1} + 3$ for $n \geq 2$.

Objective #3: Use factorial notation.

✔ *Solved Problem #3*

3. Write the first four terms of the sequence whose nth term is $a_n = \dfrac{20}{(n+1)!}$.

$$a_n = \frac{20}{(n+1)!}$$
$$a_1 = \frac{20}{(1+1)!} = \frac{20}{2!} = 10$$
$$a_2 = \frac{20}{(2+1)!} = \frac{20}{3!} = \frac{20}{6} = \frac{10}{3}$$
$$a_3 = \frac{20}{(3+1)!} = \frac{20}{4!} = \frac{20}{24} = \frac{5}{6}$$
$$a_4 = \frac{20}{(4+1)!} = \frac{20}{5!} = \frac{20}{120} = \frac{1}{6}$$

The first four terms are 10, $\frac{10}{3}$, $\frac{5}{6}$, and $\frac{1}{6}$.

Pencil Problem #3

3. Write the first four terms of the sequence whose nth term is $a_n = \dfrac{n^2}{n!}$.

Objective #4: Use summation notation.

✔ *Solved Problem #4*

4a. Expand and evaluate the sum: $\sum_{k=3}^{5}\left(2^k - 3\right)$.

$$\sum_{k=3}^{5}\left(2^k - 3\right)$$
$$= \left(2^3 - 3\right) + \left(2^4 - 3\right) + \left(2^5 - 3\right)$$
$$= (8-3) + (16-3) + (32-3)$$
$$= 5 + 13 + 29$$
$$= 47$$

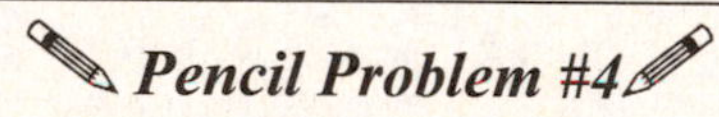

Pencil Problem #4

4a. Expand and evaluate the sum: $\sum_{k=1}^{5} k(k+4)$.

4b. Expand and evaluate the sum: $\sum_{i=1}^{5} 4$.

$$\sum_{i=1}^{5} 4 = 4+4+4+4+4$$
$$= 20$$

4b. Expand and evaluate the sum: $\sum_{i=5}^{9} 11$

4c. Express the sum using summation notation.
Use 1 as the lower limit of summation and i for the index of summation.

$$1^2+2^2+3^2+\cdots+9^2$$

The sum has nine terms, each of the form i^2, starting at $i=1$ and ending at $i=9$.

$$1^2+2^2+3^2+\cdots+9^2 = \sum_{i=1}^{9} i^2$$

4c. Express the sum using summation notation.
Use 1 as the lower limit of summation and i for the index of summation.

$$2+2^2+2^3+\ldots+2^{11}$$

4d. Express the sum using summation notation.
Use 1 as the lower limit of summation and i for the index of summation.

$$1+\frac{1}{2}+\frac{1}{4}+\frac{1}{8}+\cdots+\frac{1}{2^{n-1}}$$

The sum has n terms, each of the form $\frac{1}{2^{i-1}}$, starting at $i=1$ and ending at $i=n$.

$$1+\frac{1}{2}+\frac{1}{4}+\frac{1}{8}+\cdots+\frac{1}{2^{n-1}} = \sum_{i=1}^{n} \frac{1}{2^{i-1}}$$

4d. Express the sum using summation notation.
Use 1 as the lower limit of summation and i for the index of summation.

$$\frac{1}{2}+\frac{2}{3}+\frac{3}{4}+\ldots+\frac{14}{14+1}$$

Answers for Pencil Problems *(Textbook Exercise references in parentheses)*:

1a. 5, 8, 11, 14 *(8.1 #1)*

1b. –4, 5, –6, 7 *(8.1 #7)*

2. 4, 11, 25, 53 *(8.1 #17)*

3. $1, 2, \frac{3}{2}, \frac{2}{3}$ *(8.1 #19)*

4a. 115 *(8.1 #33)*

4b. 55 *(8.1 #37)*

4c. $\sum_{i=1}^{11} 2^i$ *(8.1 #45)*

4d. $\sum_{i=1}^{14} \frac{i}{i+1}$ *(8.1 #49)*

Section 8.2
Arithmetic Sequences

IT'S A FULL THEATER TONIGHT!

Some theaters have the same number of seats in each row. But other theaters are more fan-shaped.

In this section of the textbook, we will encounter such a fan-shaped theater, and we will use the techniques of this section to quickly determine the total number of seats without actually adding the number in each row.

Objective #1: Find the common difference for an arithmetic sequence.

✔ *Solved Problem #1*

1. True or false: An arithmetic sequence is a sequence in which each term after the first differs from the preceding term by a constant amount.

true

Pencil Problem #1

1. True or false: In an arithmetic sequence, each term after the first term can be obtained by adding the common difference to the preceding term.

Objective #2: Write terms of an arithmetic sequence.

✔ *Solved Problem #2*

2. Write the first six terms of the arithmetic sequence with first term 100 and common difference -30.

$$a_1 = 100$$
$$a_2 = 100 + (-30) = 70$$
$$a_3 = 70 + (-30) = 40$$
$$a_4 = 40 + (-30) = 10$$
$$a_5 = 10 + (-30) = -20$$
$$a_6 = -20 + (-30) = -50$$

Pencil Problem #2

2. Write the first six terms of the arithmetic sequence with first term -7 and common difference 4.

Objective #3: Use the formula for the general term of an arithmetic sequence.

✔ Solved Problem #3

3a. Find the ninth term of the arithmetic sequence whose first term is 6 and whose common difference is −5.

$a_1 = 6,\ d = -5$

To find the ninth term, a_9, replace n in the formula with 9, replace a_1 with 6, and replace d with −5.

$$
\begin{aligned}
a_n &= a_1 + (n-1)d \\
a_9 &= 6 + (9-1)(-5) \\
&= 6 + 8(-5) \\
&= 6 + (-40) \\
&= -34
\end{aligned}
$$

3b. In 2010, 16% of the U.S. population was Latino. On average, this is projected to increase by approximately 0.35% per year. Write a formula for the *n*th term of the arithmetic sequence that describes the percentage of the U.S. population that will be Latino n years after 2009.

$$
\begin{aligned}
a_n &= a_1 + (n-1)d \\
&= 16 + (n-1)0.35 \\
&= 0.35n + 15.65
\end{aligned}
$$

3c. Use the result from the previous problem to project the percentage of the U.S. population that will be Latino in 2030.

2030 is 21 years after 2009.

$a_n = 0.35n + 15.65$

$a_{20} = 0.35(21) + 15.65 = 23$

In 2030, 23% of the U.S. population is projected to be Latino.

Pencil Problem #3

3a. Find the 50th term of the arithmetic sequence whose first term is 7 and whose common difference is 5.

3b. In 1970, 11.0% of Americans ages 25 and older had completed four years of college or more. On average, this percentage has increased by approximately 0.5 each year. Write a formula for the *n*th term of the arithmetic sequence that models the percentage of Americans ages 25 and older who had or will have completed four years of college or more n years after 1969.

3c. Use the result from the previous problem to project the percentage of Americans ages 25 and older who will have completed four years of college or more by 2019.

Objective #4: Use the formula for the sum of the first n terms of an arithmetic sequence.

✔ Solved Problem #4

4a. Find the sum of the first 15 terms of the arithmetic sequence: 3, 6, 9, 12, ...

To find the sum of the first 15 terms, S_{15}, replace n in the formula with 15.

$$S_n = \frac{n}{2}(a_1 + a_n)$$
$$S_{15} = \frac{15}{2}(a_1 + a_{15})$$

Use the formula for the general term of a sequence to find a_{15}. The common difference, d, is 3, and the first term, a_1, is 3.

$$a_n = a_1 + (n-1)d$$
$$a_{15} = 3 + (15-1)(3)$$
$$= 3 + 14(3)$$
$$= 3 + 42$$
$$= 45$$

Thus, $S_{15} = \frac{15}{2}(3+45) = \frac{15}{2}(48) = 360$.

4b. Find the following sum: $\sum_{i=1}^{30}(6i-11)$.

$$\sum_{i=1}^{30}(6i-11)$$
$$= (6\cdot1-11) + (6\cdot2-11) + (6\cdot3-11) + \ldots + (6\cdot30-11)$$
$$= -5 + 1 + 7 + \ldots + 169$$

The first term, a_1, is –5.
The common difference, d, is $1-(-5)=6$.
The last term, a_{30}, is 169.

$$S_n = \frac{n}{2}(a_1 + a_n)$$
$$S_{30} = \frac{30}{2}(-5+169)$$
$$= 15(164)$$
$$= 2460$$

Thus, $\sum_{i=1}^{30}(6i-11) = 2460$

Pencil Problem #4

4a. Find the sum of the first 50 terms of the arithmetic sequence: $-10, -6, -2, 2, \ldots$

4b. Find the following sum: $\sum_{i=1}^{100} 4i$.

4c. The model $a_n = 1800n + 64{,}130$ describes yearly adult residential community costs n years after 2013. How much would it cost for the adult residential community for a ten-year period beginning in 2014?

$a_n = 1800n + 64{,}130$
$a_1 = 1800(1) + 64{,}130 = 65{,}930$
$a_{10} = 1800(10) + 64{,}130 = 82{,}130$

$S_n = \frac{n}{2}(a_1 + a_n)$
$S_{10} = \frac{10}{2}(a_1 + a_{10})$
$= 5(65{,}930 + 82{,}130)$
$= 5(148{,}060)$
$= \$740{,}300$

It would cost $740,300 for the ten-year period beginning in 2014.

4c. A section in a stadium has 20 seats in the first row, 23 seats in the second row, increasing by 3 seats each row for a total of 38 rows. How many seats are in this section of the stadium?

Answers for Pencil Problems *(Textbook Exercise references in parentheses)*:

1. true *(8.2 #1)* **2.** $-7, -3, 1, 5, 9, 13$ *(8.2 #3)*

3a. 252 *(8.2 #17)* **3b.** $a_n = 0.5n + 10.5$ *(8.2 #61a)* **3c.** 35.5% *(8.2 #61b)*

4a. 4400 *(8.2 #37)* **4b.** 20,200 *(8.2 #49)* **4c.** 2869 seats *(8.2 #71)*

Section 8.3
Geometric Sequences and Series

How Much Will You End Up With?

Suppose you are 24 and you have just landed a job!

You decide that you can save for retirement by putting aside \$80 per month into an account which pays 5% compounded monthly.

What will the account balance be when you reach age 65?

In this section, several applications will deal with money and we will apply geometric sequences and series to find answers to a variety of financial questions.

Objective #1: Find the common ratio of a geometric sequence.

✔ Solved Problem #1

1. True or False: The sequence
$$6, -12, 24, -48, 96, ...$$
is an example of a geometric sequence.

True. Each term after the first term is -2 times the previous term. The common ration is -2.

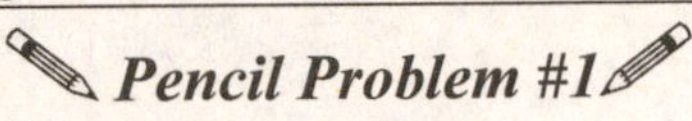

1. True or False: The sequence
$$2, 6, 24, 120, ...$$
is an example of a geometric sequence.

Objective #2: Write terms of a geometric sequence.

✔ Solved Problem #2

2. Write the first six terms of the geometric sequence with first term 12 and common ratio $\frac{1}{2}$.

$a_1 = 12,\ r = \dfrac{1}{2}$

$a_2 = 12\left(\frac{1}{2}\right)^1 = 6$

$a_3 = 12\left(\frac{1}{2}\right)^2 = \frac{12}{4} = 3$

$a_4 = 12\left(\frac{1}{2}\right)^3 = \frac{12}{8} = \frac{3}{2}$

$a_5 = 12\left(\frac{1}{2}\right)^4 = \frac{12}{16} = \frac{3}{4}$

$a_6 = 12\left(\frac{1}{2}\right)^5 = \frac{12}{32} = \frac{3}{8}$

The first six terms are $12, 6, 3, \frac{3}{2}, \frac{3}{4}$, and $\frac{3}{8}$.

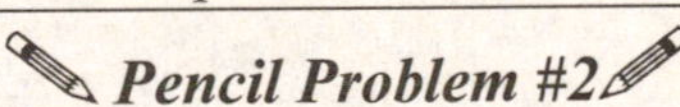

2. Write the first five terms of the geometric sequence with first term 5 and common ratio 3.

Objective #3: Use the formula for the general term of a geometric sequence.

✔ Solved Problem #3

3a. Find the seventh term of the geometric sequence whose first term is 5 and whose common ratio is −3.

$a_1 = 5, r = -3$

$a_n = a_1 r^{n-1}$

$a_7 = 5(-3)^{7-1}$

$= 5(-3)^6$

$= 5(729)$

$= 3645$

The seventh term is 3645.

3b. Write the general term for the geometric sequence: $3, 6, 12, 24, 48, \ldots$
Then use the formula for the general term to find the eighth term.

$r = \frac{6}{3} = 2,\ a_1 = 3$

Formula for the general term:

$a_n = a_1(r)^{n-1}$

$a_n = 3(2)^{n-1}$

Find the eighth term:

$a_n = 3(2)^{n-1}$

$a_8 = 3(2)^{8-1}$

$= 3(2)^7$

$= 3(128)$

$= 384$

The eighth term is 384.

Pencil Problem #3

3a. Find the eighth term of the geometric sequence whose first term is 6 and whose common ratio is 2.

3b. Write the general term for the geometric sequence: $3, 12, 48, 192, \ldots$
Then use the formula for the general term to find the seventh term.

Objective #4: Use the formula for the sum of the first n terms of a geometric sequence.

Solved Problem #4

4a. Find the sum of the first nine terms of the geometric sequence: $2, -6, 18, -54, \ldots$

$a_1 = 2,\ r = \dfrac{-6}{2} = -3$

$S_n = \dfrac{a_1(1-r^n)}{1-r}$

$S_9 = \dfrac{2\left(1-(-3)^9\right)}{1-(-3)} = \dfrac{2(19{,}684)}{4} = 9842$

The sum of the first nine terms is 9842.

4b. Find the following sum: $\sum_{i=1}^{8} 2 \cdot 3^i$.

$a_1 = 2 \cdot (3)^1 = 6,\ r = 3$

$S_n = \dfrac{a_1(1-r^n)}{1-r}$

$S_8 = \dfrac{6\left(1-3^8\right)}{1-3} = \dfrac{6(-6560)}{-2} = 19{,}680$

Thus, $\sum_{i=1}^{8} 2 \cdot 3^i = 19{,}680$.

Pencil Problem #4

4a. Find the sum of the first 11 terms of the geometric sequence: $3, -6, 12, -24, \ldots$

4b. Find the following sum: $\sum_{i=1}^{10} 5 \cdot 2^i$.

4c. A job pays a salary of \$30,000 the first year. During the next 29 years, the salary increases by 6% each year. What is the total lifetime salary over the 30-year period? Round to the nearest dollar.

$$a_1 = 30{,}000,\ r = 1.06$$

$$S_n = \frac{a_1(1-r^n)}{1-r}$$

$$S_{30} = \frac{30{,}000\left(1-(1.06)^{30}\right)}{1-1.06} \approx 2{,}371{,}746$$

The total lifetime salary is \$2,371,746.

4c. A job pays a salary of \$24,000 the first year. During the next 19 years, the salary increases by 5% each year. What is the total lifetime salary over the 20-year period? Round to the nearest dollar.

Objective #5: Find the value of an annuity.

✔ *Solved Problem #5*

5. At age 30, to save for retirement, you decide to deposit \$100 at the end of each month into an IRA that pays 9.5% compounded monthly. Find how much will you have in the IRA when you retire at age 65 and find how much is interest.

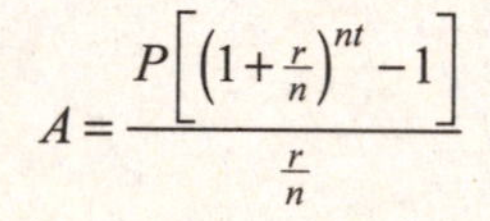

$$A = \frac{P\left[\left(1+\frac{r}{n}\right)^{nt}-1\right]}{\frac{r}{n}}$$

$$P = 100,\ r = 0.095,\ n = 12,\ t = 35$$

$$A = \frac{100\left[\left(1+\frac{0.095}{12}\right)^{12\cdot 35}-1\right]}{\frac{0.095}{12}} \approx 333{,}946$$

The value of the IRA will be \$333,946.

Find the interest:

$$\begin{aligned}\text{Interest} &= \text{Value of IRA} - \text{Total deposits}\\ &\approx \$333{,}946 - \$100\cdot 12\cdot 35\\ &\approx \$333{,}946 - \$42{,}000\\ &\approx \$291{,}946\end{aligned}$$

Pencil Problem #5

5. At age 25, to save for retirement, you decide to deposit \$50 at the end of each month into an IRA that pays 5.5% compounded monthly. Find how much will you have in the IRA when you retire at age 65 and find how much is interest.

Objective #6: Use the formula for the sum of an infinite geometric series.

✔ *Solved Problem #6*

6a. Find the sum of the infinite geometric series:

$$3+2+\frac{4}{3}+\frac{8}{9}+\cdots$$

$a_1 = 3,\ r = \frac{2}{3}$

$$S = \frac{a_1}{1-r}$$

$$S = \frac{3}{1-\frac{2}{3}}$$

$$= \frac{3}{\frac{1}{3}}$$

$$= 9$$

The sum of this infinite geometric series is 9.

6b. Express $0.\overline{9}$ as a fraction in lowest terms.

$$0.\overline{9} = 0.9999\cdots = \frac{9}{10}+\frac{9}{100}+\frac{9}{1000}+\cdots$$

$a_1 = \frac{9}{10},\ r = \frac{1}{10}$

$$S = \frac{a_1}{1-r}$$

$$S = \frac{\frac{9}{10}}{1-\frac{1}{10}}$$

$$= \frac{\frac{9}{10}}{\frac{9}{10}}$$

$$= 1$$

An equivalent fraction for $0.\overline{9}$ is 1.

 *Pencil Problem #6*

6a. Find the sum of the infinite geometric series:

$$3+\frac{3}{4}+\frac{3}{4^2}+\frac{3}{4^3}+\cdots$$

6b. Express $0.\overline{5}$ as a fraction in lowest terms.

Answers for Pencil Problems *(Textbook Exercise references in parentheses)*:

1. False *(8.3 #105)*

2. 5,15,45,135,405 *(8.3 #1)*

3a. 768 *(8.3 #9)* **3b.** $a_n = 3(4)^{n-1}$; $a_7 = 12{,}288$ *(8.3 #17)*

4a. 2049 *(8.3 #29)* **4b.** 10,230 *(8.3 #33)* **4c.** $793,583 *(8.3 #73)*

5. $87,052; $63,052 *(8.3 #79)*

6a. 4 *(8.3 #39)* **6b.** $\frac{5}{9}$ *(8.3 #45)*

Section 8.4
Mathematical Induction

Will They ALL Fall Down?

The mathematical principle of this section can be illustrated using an unending line of dominoes. If the first domino is pushed over, it knocks down the next, which knocks down the next, and so on, in a chain reaction.

To topple all the dominoes in the infinite sequence, two conditions must be satisfied:

1. The first domino must be knocked down.
2. If the domino in position k is knocked down, then the domino in position $k+1$ must be knocked down.

If the second condition is not satisfied, it does not follow that all the dominoes will topple. For example, suppose the dominoes are spaced far enough apart so that a falling domino does not push over the next domino in the line.

Objective #1: Understand the principle of mathematical induction.

✔ Solved Problem #1a

1a. For the given statement S_n, write the statement S_1.

$$S_n: 2+4+6+\cdots+2n = n(n+1)$$

If $n=1$ then the statement S_1 is obtained by writing the first term, 2, on the left, and substituting 1 for n on the right.

$$S_1: 2 = 1(1+1)$$

Pencil Problem #1a

1a. For the given statement S_n, write the statement S_1.

$$S_n: 1+3+5+(2n-1) = n^2$$

✔ *Solved Problem #1b*

1b. For the given statement S_n, write the two statements S_k, and S_{k+1}.

$$S_n : 1^3 + 2^3 + 3^3 + \cdots + n^3 = \frac{n^2(n+1)^2}{4}$$

Write S_k by taking the sum of the first k terms on the left and replacing n with k on the right.

$$S_k : 1^3 + 2^3 + 3^3 + \cdots + k^3 = \frac{k^2(k+1)^2}{4}$$

Write S_{k+1} by taking the sum of the first $k+1$ terms on the left and replacing n with $k+1$ on the right.

$$S_{k+1} : 1^3 + 2^3 + 3^3 + \cdots + k^3 + (k+1)^3 = \frac{(k+1)^2(k+1+1)^2}{4}$$

$$= \frac{(k+1)^2(k+2)^2}{4}$$

✎ *Pencil Problem #1b* ✎

1b. For the given statement S_n, write the two statements S_k, and S_{k+1}.

$$S_n : 3 + 7 + 11 + \cdots + (4n-1) = n(2n+1)$$

Objective #2: Prove statements using mathematical induction.

✔ ***Solved Problem #2***

2. Use mathematical induction to prove that $1^3+2^3+3^3+\cdots+n^3=\dfrac{n^2(n+1)^2}{4}$ for all positive integers n.

Step 1. *Show that* S_1 *is true*:

$$1^3=\frac{1^2(1+1)^2}{4}$$
$$1=\frac{1(2)^2}{4}$$
$$1=\frac{4}{4}$$
$$1=1,\ \text{True}$$

Step 2. *Show that if* S_k *is true, then* S_{k+1} *is true*:

Assume $1^3+2^3+3^3+\cdots+k^3=\dfrac{k^2(k+1)^2}{4}$ is true. Then,

$$1^3+2^3+3^3+\cdots+k^3+(k+1)^3=\frac{k^2(k+1)^2}{4}+(k+1)^3$$
$$1^3+2^3+3^3+\cdots+k^3+(k+1)^3=\frac{k^2(k+1)^2}{4}+\frac{4(k+1)^3}{4}$$
$$1^3+2^3+3^3+\cdots+k^3+(k+1)^3=\frac{k^2(k+1)^2+4(k+1)^3}{4}$$
$$1^3+2^3+3^3+\cdots+k^3+(k+1)^3=\frac{(k+1)^2\left(k^2+4(k+1)\right)}{4}$$
$$1^3+2^3+3^3+\cdots+k^3+(k+1)^3=\frac{(k+1)^2\left(k^2+4k+4\right)}{4}$$
$$1^3+2^3+3^3+\cdots+k^3+(k+1)^3=\frac{(k+1)^2(k+2)^2}{4}$$

The final statement is S_{k+1}.

Thus, by mathematical induction, the statement $1^3+2^3+3^3+\cdots+n^3=\dfrac{n^2(n+1)^2}{4}$ is true for all positive integers n.

Pencil Problem #2

2. Use mathematical induction to prove that $1+2+2^2+\cdots+2^{n-1}=2^n-1$ for all positive integers n.

Answers for Pencil Problems *(Textbook Exercise references in parentheses)*:

1a. $S_1: 1=1^2$ *(8.4 #1)*

1b. $S_k: 3+7+11+\cdots+(4k-1)=k(2k+1)$

$S_{k+1}: 3+7+11+\cdots+[4(k+1)-1]=(k+1)[2(k+1)+1]=(k+1)(2k+3)$ *(8.4 #7)*

2. *Show that* S_1 *is true*: $1=2^1-1$

$1=2-1$

$1=1$, true

Show that if S_k *is true, then* S_{k+1} *is true*:

Assume $S_k: 1+2+2^2+\cdots+2^{k-1}=2^k-1$ is true. Then, $1+2+2^2+\cdots+2^{k-1}+2^{(k+1)-1}=2^k-1+2^{(k+1)-1}$

$$1+2+2^2+\cdots+2^{k-1}+2^{(k+1)-1}=2^k-1+2^k$$

$$1+2+2^2+\cdots+2^{k-1}+2^{(k+1)-1}=2\cdot 2^k-1$$

$$1+2+2^2+\cdots+2^{k-1}+2^{(k+1)-1}=2^{k+1}-1$$

Thus, $1+2+2^2+\cdots+2^{n-1}=2^n-1$ is true for all positive integers n. *(8.4 #17)*

Section 8.5
The Binomial Theorem

Who Knew That First?

Telephones, Internet, and other modern forms of communication mean that information now can spread across the globe in the blink of an eye.

In this section, we study a special array of numbers known as Pascal's triangle, named after mathematician Blaise Pascal.
However, this triangular array of numbers actually appeared centuries earlier in a Chinese document.

The same mathematics is often discovered by independent researchers separated by time, place, and culture.
But with modern communication, important discoveries are now shared much more efficiently.

Objective #1: Evaluate a binomial coefficient.

✔ *Solved Problem #1*

1a. Evaluate: $\binom{6}{3}$.

$$\binom{6}{3}=\frac{6!}{3!(6-3)!}$$
$$=\frac{6!}{3!3!}$$
$$=\frac{6\cdot5\cdot4\cdot\cancel{3!}}{3\cdot2\cdot1\cdot\cancel{3!}}$$
$$=20$$

1b. Evaluate: $\binom{6}{0}$.

$$\binom{6}{0}=\frac{6!}{0!(6-0)!}$$
$$=\frac{6!}{6!}$$
$$=1$$

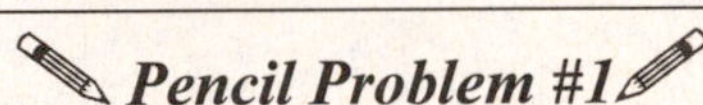

Pencil Problem #1

1a. Evaluate: $\binom{8}{3}$.

1b. Evaluate: $\binom{12}{1}$.

1c. Evaluate: $\binom{8}{2}$.

$$\binom{8}{2} = \frac{8!}{2!(8-2)!} = \frac{8!}{2!6!} = \frac{8 \cdot 7}{2} = 28$$

1c. Evaluate: $\binom{100}{2}$.

1d. Evaluate: $\binom{3}{3}$.

$$\binom{3}{3} = \frac{3!}{3!(3-3)!} = \frac{3!}{3!0!} = \frac{3!}{3!} = 1$$

1d. Evaluate: $\binom{6}{6}$.

Objective #2: Expand a binomial raised to a power.

✔ ***Solved Problem #2a***

2a. Expand: $(x+1)^4$

$$(x+1)^4 = \binom{4}{0}x^4 + \binom{4}{1}x^3 + \binom{4}{2}x^2 + \binom{4}{3}x + \binom{4}{4} = x^4 + 4x^3 + 6x^2 + 4x + 1$$

Pencil Problem #2a

2a. Expand: $(x+2)^3$

✔ *Solved Problem #2b*

2b. Expand: $(x-2y)^5$

$$(x-2y)^5 = \binom{5}{0}x^5(-2y)^0 + \binom{5}{1}x^4(-2y)^1 + \binom{5}{2}x^3(-2y)^2 + \binom{5}{3}x^2(-2y)^3 + \binom{5}{4}x(-2y)^4 + \binom{5}{5}x^0(-2y)^5$$
$$= x^5 \quad -5x^4(2y) \quad +10x^3(4y^2) \quad -10x^2(8y^3) \quad +5x(16y^4) \quad -32y^5$$
$$= x^5 - 10x^4y + 40x^3y^2 - 80x^2y^3 + 80xy^4 - 32y^5$$

Pencil Problem #2b

2b. Expand: $(x^2+2y)^4$

Objective #3: Find a particular term in a binomial expansion.

✔ *Solved Problem #3*

3. Find the fifth term in the expansion of $(2x+y)^9$.

Since we are looking for the 5th term, $r=5-1=4$.
Thus, $r=4$, $a=2x$, $b=y$, and $n=9$.

$$(r+1)\text{st term} = \binom{n}{r}a^{n-r}b^r$$

$$\text{fifth term} = \binom{9}{4}(2x)^5y^4$$

$$= \frac{9!}{4!5!}(32x^5)y^4$$

$$= 4032x^5y^4$$

Pencil Problem #3

3. Find the sixth term in the expansion of $\left(x^2+y^3\right)^8$.

Answers for Pencil Problems *(Textbook Exercise references in parentheses)*:

1a. 56 *(8.5 #1)*

1b. 12 *(8.5 #3)*

1c. 4950 *(8.5 #7)*

1d. 1 *(8.5 #5)*

2a. $x^3+6x^2+12x+8$ *(8.5 #9)*

2b. $x^8+8x^6y+24x^4y^2+32x^2y^3+16y^4$ *(8.5 #17)*

3. $56x^6y^{15}$ *(8.5 #43)*

Section 8.6
Counting Principles, Permutations, and Combinations

I Have NOTHING to Wear!

On many mornings, we feel quite limited on the fashion statement we desire to make that day. But the truth is we usually have more clothing options than we might think. If we consider how the various components of our outfit can be mixed and matched, the number of unique outfits can be difficult to count.

Attempting to count each possibility one-by-one can be daunting in many situations.

In this section of the textbook, we will use organized mathematical methods and formulas that will allow us to count more quickly and accurately than the 1, 2, 3, 4, 5, 6, … method.

Objective #1: Use the Fundamental Counting Principle.

✔ *Solved Problem #1*

1a. A pizza can be ordered with three choices of size (small, medium, or large), four choices of crust (thin, thick, crispy, or regular), and six choices of toppings (ground beef, sausage, pepperoni, bacon, mushrooms, or onions). How many different one-topping pizzas can be ordered?

Multiply the number of choices for each of the three decisions:

$$\underset{3}{\underline{\text{Size}}}: \times \underset{4}{\underline{\text{Crust}}}: \times \underset{6}{\underline{\text{Topping}}}: = 72$$

72 different one-topping pizzas can be ordered.

1b. License plates in a particular state display two letters followed by three numbers, such as AT-887 or BB-013. How many different license plates can be manufactured?

Multiply the number of choices for each of the letters and each of the digits:

$$\overbrace{26}^{\text{Letter 1}} \times \overbrace{26}^{\text{Letter 2}} \times \overbrace{10}^{\text{Digit 1}} \times \overbrace{10}^{\text{Digit 2}} \times \overbrace{10}^{\text{Digit 3}} = 676{,}000$$

676,000 different license plates can be manufactured.

Pencil Problem #1

1a. An ice cream store sells two drinks (sodas or milk shakes), in four sizes (small, medium, large, or jumbo), and five flavors (vanilla, strawberry, chocolate, coffee, or pistachio). In how many ways can a customer order a drink?

1b. You are taking a multiple-choice test that has five questions. Each of the questions has three answer choices, with one correct answer per question. If you select one of these three choices for each question and leave nothing blank, in how many ways can you answer the questions?

Objective #2: Use the permutations formula.

✔ *Solved Problem #2*

2a. A corporation has seven members on its board of directors. In how many different ways can it elect a president, vice-president, secretary, and treasurer?

The corporation is choosing 4 officers from a group of 7 people. The order in which the officers are chosen matters because the president, vice-president, secretary, and treasurer each have different responsibilities. Thus, we are looking for the number of permutations of 7 things taken 4 at a time.

$${}_7P_4 = \frac{7!}{(7-4)!} = \frac{7!}{3!} = 840$$

There are 840 ways of filling the four offices.

2b. In how many ways can 6 books be lined up along a shelf?

Because you are using all six of your books in every possible arrangement, you are arranging 6 books from a group of 6 books. Thus, we are looking for the number of permutations of 6 things taken 6 at a time.

$${}_6P_6 = \frac{6!}{(6-6)!} = \frac{6!}{0!} = \frac{6\cdot 5\cdot 4\cdot 3\cdot 2\cdot 1}{1} = 720$$

There are 720 ways the 6 books can be lined up along the shelf.

Pencil Problem #2

2a. Using 15 flavors of ice cream, how many cones with three different flavors can you create if it is important to you which flavor goes on the top, middle, and bottom?

2b. What is the number of permutations of 8 things taken 0 at a time?

Objective #3: Distinguish between permutation problems and combination problems.

✔ *Solved Problem #3*

3a. Determine if the question involves combinations or permutations. (Do *not* solve the problem.)

How many ways can you select 6 free DVDs from a list of 200 DVDs?

The order in which the DVDs are selected does not matter.

Thus, this problem involves combinations.

Pencil Problem #3

3a. Determine if the question involves combinations or permutations. (Do *not* solve the problem.)

A medical researcher needs 6 people to test the effectiveness of an experimental drug. If 13 people have volunteered for the test, in how many ways can 6 people be selected?

3b. Determine if the question involves combinations or permutations. (Do *not* solve the problem.)

In a race in which there are 50 runners and no ties, in how many ways can the first three finishers come in?

The order in which the runners finish does matter.

Thus, this problem involves permutations.

3b. Determine if the question involves combinations or permutations. (Do *not* solve the problem.)

How many different four-letter passwords can be formed from the letters A, B, C, D, E, F, and G if no repetition of letters is allowed?

Objective #4: Use the combinations formula.

 ***Solved Problem #4***

4a. From a group of 10 physicians, in how many ways can four people be selected to attend a conference on acupuncture?

The order in which the four people are selected does not matter. This is a problem of selecting 4 people from a group of 10 people. We are looking for the number of combinations of 10 things taken 4 at a time.

$$\begin{aligned} {}_{10}C_4 &= \frac{10!}{(10-4)!4!} \\ &= \frac{10!}{6!4!} \\ &= \frac{10\cdot 9\cdot 8\cdot 7\cdot 6!}{6!\cdot 4\cdot 3\cdot 2\cdot 1} \\ &= 210 \end{aligned}$$

The four attendees can be selected in 210 different ways.

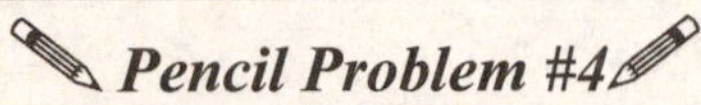

Pencil Problem #4

4a. An election ballot asks voters to select three city commissioners from a group of six candidates. In how many ways can this be done?

4b. How many different 4-card hands can be dealt from a deck that has 16 different cards?

Because the order in which the 4 cards are dealt does not matter, this is a problem involving combinations. We are looking for the number of combinations of 16 cards drawn 4 at a time.

$$\begin{aligned} {}_{16}C_4 &= \frac{16!}{(16-4)!4!} \\ &= \frac{16!}{12!4!} \\ &= \frac{16\cdot 15\cdot 14\cdot 13\cdot 12!}{12!\cdot 4\cdot 3\cdot 2\cdot 1} \\ &= \frac{16\cdot 15\cdot 14\cdot 13\cdot \cancel{12!}}{\cancel{12!}\cdot 4\cdot 3\cdot 2\cdot 1} \\ &= \frac{16\cdot 15\cdot 14\cdot 13}{4\cdot 3\cdot 2\cdot 1} \\ &= 1820 \end{aligned}$$

There are 1820 different 4-card hands.

4b. You volunteer to help drive children at a charity event to the zoo, but you can fit only 8 of the 17 children present in your van. How many different groups of 8 children can you drive?

Answers for Pencil Problems *(Textbook Exercise references in parentheses)*:

1a. 40 *(8.6 #31)* **1b.** 243 *(8.6 #33)*

2a. 2730 *(8.6 #65)* **2b.** 1 *(8.6 #7)*

3a. combinations *(8.6 #17)* **3b.** permutations *(8.6 #19)*

4a. 20 *(8.6 #49)* **4b.** 24,310 *(8.6 #53)*

Section 8.7
Probability

The Weather Outside is Frightful!

Have you ever thought about the chances of being hit by lightning, caught in a tornado, hurricane, or some other major weather event?

In one of the application exercises in this section, mathematicians, meteorologists, and you will team up to determine such probabilities.

Objective #1: Compute empirical probability.

✔ *Solved Problem #1*

1. Use the data in the table to find the probabilities.

Mammography Screening on 100,000 U.S. Women, Ages 40 to 50	Breast Cancer	No Breast Cancer
Positive Mammogram	720	6944
Negative Mammogram	80	92,256

1a. Find the probability that a woman aged 40 to 50 has a positive mammogram.

The probability of having a positive mammogram is the number of women with a positive mammogram divided by the total number of women.

$$
\begin{aligned}
P(\text{positive mammogram}) &= \frac{720+6944}{100{,}000} \\
&= \frac{7664}{100{,}000} \\
&\approx 0.077
\end{aligned}
$$

Pencil Problem #1

1. The table shows the distribution, by marital status and gender, of the 242 million Americans ages 18 or older. Use the table to find the probabilities.

	Never Married	Married	Widowed	Divorced
Male	40	65	3	10
Female	34	65	11	14

1a. If one person is randomly selected from the population described in the table, find the probability, to the nearest hundredth, that the person is divorced.

1b. Among women with positive mammograms, find the probability of having breast cancer. (Use the data in the table on the previous page)

To find the probability of breast cancer among women with positive mammograms, restrict the data to women with positive mammograms:

Mammography Screening on 100,000 U.S. Women, Ages 40 to 50	Breast Cancer	No Breast Cancer
Positive Mammogram	720	6944

$$P(\text{breast cancer}) = \frac{720}{720+6944} = \frac{720}{7664} \approx 0.094$$

1b. Among those who are divorced, find the probability of selecting a woman. (Use the data in the table on the previous page)

Objective #2: Compute theoretical probability.

✔ *Solved Problem #2*

2a. A die is rolled. Find the probability of getting a number greater than 4.

Two of the six numbers, 5 and 6, are greater than 4.

$$P(\text{greater than } 4) = \frac{2}{6} = \frac{1}{3}$$

2b. The original Florida LOTTO was set up so that each player chose six different numbers from 1 to 49. With one LOTTO ticket, what was the probability of winning the top cash prize? Express the answer as a fraction and as a decimal correct to ten places.

$$_{49}C_6 = \frac{49!}{(49-6)!6!} = \frac{49!}{43!6!} = \frac{49\cdot 48\cdot 47\cdot 46\cdot 45\cdot 44\cdot \cancel{43!}}{\cancel{43!}\cdot 6\cdot 5\cdot 4\cdot 3\cdot 2\cdot 1} = 13{,}983{,}816$$

$$P(\text{winning LOTTO}) = \frac{1}{13{,}983{,}816} \approx 0.0000000715$$

Pencil Problem #2

2a. A die is rolled. Find the probability of getting a 4.

2b. To play the California lottery, a person has to correctly select 6 out of 51 numbers. If you pick six numbers that are the same as the ones drawn by the lottery, you win. What is the probability that a person with one combination of six numbers will win?

Objective #3: Find the probability that an event will not occur.

✔ *Solved Problem #3*

3. Of the 7000 million people in the world, 550 million live in North America. If one person is randomly selected from the world population, find the probability that the person does not live in North America.

$$
\begin{aligned}
P(\text{not North America}) &= 1 - P(\text{North America}) \\
&= 1 - \frac{550}{7000} \\
&= \frac{6450}{7000} \\
&= \frac{129}{140}
\end{aligned}
$$

Pencil Problem #3

3. If you are dealt one card from a 52-card deck, find the probability that you are *not* dealt a king.

Objective #4: Find the probability of one event or a second event occurring.

✔ *Solved Problem #4*

4a. If you roll a single, six-sided die, what is the probability of getting either a 4 or a 5?

These events are mutually exclusive.
Thus, add their individual probabilities.

$$
\begin{aligned}
P(4 \text{ or } 5) &= P(4) + P(5) \\
&= \frac{1}{6} + \frac{1}{6} \\
&= \frac{2}{6} \\
&= \frac{1}{3}
\end{aligned}
$$

4b. Each number, 1 through 8, is written on slips of paper and placed in a hat. If one number is selected at random, find the probability that the number selected will be an odd number or a number less than 5.

These events are *not* mutually exclusive. Thus, use the formula $P(A \text{ or } B) = P(A) + P(B) - P(A \text{ and } B)$.

$$
\begin{aligned}
&P(\text{odd or less than 5}) \\
&= P(\text{odd}) + P(\text{less than 5}) - P(\text{odd and less than 5}) \\
&= \frac{4}{8} + \frac{4}{8} - \frac{2}{8} \\
&= \frac{6}{8} \\
&= \frac{3}{4}
\end{aligned}
$$

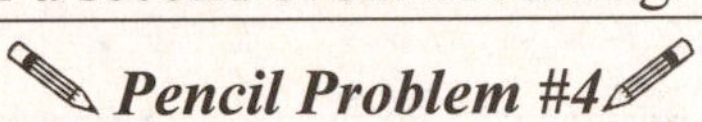

Pencil Problem #4

4a. If you are dealt one card from a 52-card deck, find the probability that you are dealt a 2 or a 3.

4b. Each number, 1 through 8, is written on slips of paper and placed in a hat. If one number is selected at random, find the probability that the number selected will be an odd number or a number less than 6.

Objective #5: Find the probability of one event and a second event occurring.

✔ Solved Problem #5

5a. On a roulette wheel, the ball can land with equal probability on any one of the 38 numbered slots, two of which are green. Find the probability of green occurring on two consecutive plays.

The events are independent.
Thus, use the formula $P(A \text{ and } B) = P(A) \cdot P(B)$.

$$P(\text{green and green}) = P(\text{green}) \cdot P(\text{green})$$
$$= \frac{2}{38} \cdot \frac{2}{38}$$
$$= \frac{1}{361}$$
$$\approx 0.00277$$

5b. Find the probability of a family having four boys in a row.

The events are independent.
Thus, multiply their probabilities.

$$P(4 \text{ boys in a row}) = P(\text{boy and boy and boy and boy})$$
$$= P(\text{boy}) \cdot P(\text{boy}) \cdot P(\text{boy}) \cdot P(\text{boy})$$
$$= \frac{1}{2} \cdot \frac{1}{2} \cdot \frac{1}{2} \cdot \frac{1}{2}$$
$$= \frac{1}{16}$$

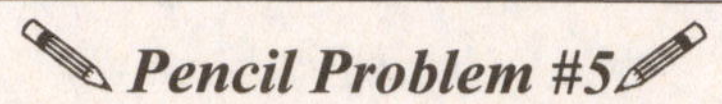

Pencil Problem #5

5a. A single die is rolled twice. Find the probability of rolling a 2 the first time and a 3 the second time.

5b. If you toss a fair coin six times, what is the probability of getting all heads?

Answers for Pencil Problems *(Textbook Exercise references in parentheses)*:

1a. 0.10 *(8.7 #1)* **1b.** 0.58 *(8.7 #7)* **2a.** $\frac{1}{6}$ *(8.7 #11)* **2b.** $\frac{1}{18{,}009{,}460} \approx 0.0000000555$ *(8.7 #27)*

3. $\frac{12}{13}$ *(8.7 #37)* **4a.** $\frac{2}{13}$ *(8.7 #39)* **4b.** $\frac{3}{4}$ *(8.7 #43)* **5a.** $\frac{1}{36}$ *(8.7 #47)* **5b.** $\frac{1}{64}$ *(8.7 #51)*

Integrated Review Worksheets
College Algebra with Integrated Review

Table of Contents

Chapter 1
Equations and Inequalities

Guided Practice:

☐ Review each of the following ***Solved Problems*** and complete each ***Pencil Problem***.

Objective 1.R.1: Evaluate algebraic expressions.

✔ Solved Problem #1

1a. Evaluate the expression $2(x+6)$ for $x=10$.

$$\begin{aligned} 2(x+6) &= 2(\underbrace{10}_{x}+6) \\ &= 2(16) \\ &= 32 \end{aligned}$$

1b. Evaluate the expression $\dfrac{6x-y}{2y-x-8}$ for $x=3$ and $y=8$.

$$\begin{aligned} \frac{6x-y}{2y-x-8} &= \frac{6\cdot\overset{x}{3}-\overset{y}{8}}{2\cdot\underset{y}{8}-\underset{x}{3}-8} \\ &= \frac{18-8}{16-3-8} \\ &= \frac{10}{5} \\ &= 2 \end{aligned}$$

Pencil Problem #1

1a. Evaluate the expression $5+3x$ for $x=4$.

1b. Evaluate the expression $\dfrac{2x-y+6}{2y-x}$ for $x=7$ and $y=5$.

Objective 1.R.2: Find coordinates of points in the rectangular coordinate system.

✔ Solved Problem #2

2. Determine the coordinates of points E, F, and G.

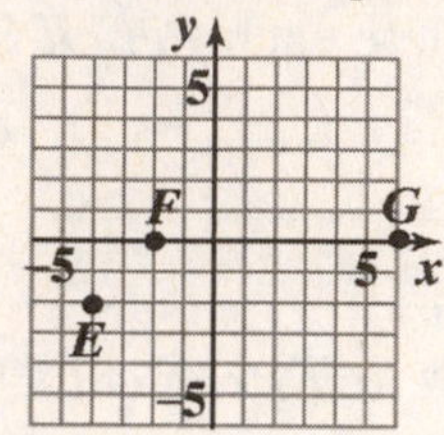

From the origin, point E is left 4 units and down 2 units.
Coordinates: $E(-4,-2)$

From the origin, point F is left 2 units.
Coordinates: $F(-2,0)$

From the origin, point G is right 6 units.
Coordinates: $G(6,0)$

Pencil Problem #2

2. Determine the coordinates of points A, C, and E.

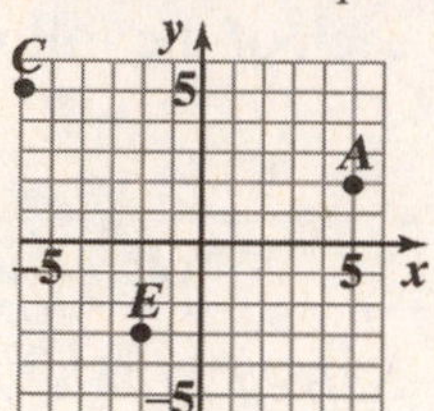

Objective 1.R.3: Find solutions of an equation in two variables.

✔ Solved Problem #3

3. Find five solutions of $y = 3x + 2$.
Select integers for x, starting with -2 and ending with 2.

x	$y = 3x + 2$	(x, y)
-2	$y = 3(-2)+2$ $= -6+2$ $= -4$	$(-2,-4)$
-1	$y = 3(-1)+2$ $= -3+2$ $= -1$	$(-1,-1)$
0	$y = 3(0)+2$ $= 0+2$ $= 2$	$(0,2)$
1	$y = 3(1)+2$ $= 3+2$ $= 5$	$(1,5)$
2	$y = 3(2)+2$ $= 6+2$ $= 8$	$(2,8)$

Pencil Problem #3

3. Find five solutions of $y = -3x + 7$.
Select integers for x, starting with -2 and ending with 2.

Objective 1.R.4: Determine whether a number is a solution of an equation.

✔ Solved Problem #4

4a. Determine whether the given number is a solution of the equation.
$9x-3=42;\ 6$

$$\begin{aligned} 9x-3&=42\\ 9(6)-3&=42\\ 54-3&=42\\ 51&=42,\ \text{false} \end{aligned}$$

6 is not a solution.

4b. Determine whether the given number is a solution of the equation.
$2(y+3)=5y-3;\ 3$

$$\begin{aligned} 2(y+3)&=5y-3\\ 2(3+3)&=5(3)-3\\ 2(6)&=15-3\\ 12&=12,\ \text{true} \end{aligned}$$

3 is a solution.

Pencil Problem #4

4a. Determine whether the given number is a solution of the equation.
$5a-4=2a+5;\ 3$

4b. Determine whether the given number is a solution of the equation.
$2(w+1)=3(w-1);\ 7$

Objective 1.R.5: Simplify algebraic expressions.

✔ Solved Problem #5

5a. Simplify: $7(2x+3)+11x$

$$\begin{aligned} 7(2x+3)+11x&=7\cdot 2x+7\cdot 3+11x\\ &=14x+21+11x\\ &=(14x+11x)+21\\ &=25x+21 \end{aligned}$$

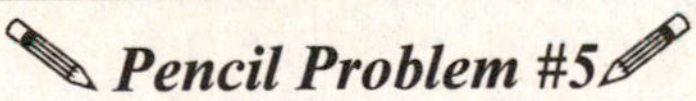

Pencil Problem #5

5a. Simplify: $5(3x+2)-4$

5b. Simplify: $7(4x+3y)+2(5x+y)$

$$\begin{aligned} 7(4x+3y)+2(5x+y) &= 7\cdot 4x+7\cdot 3y+2\cdot 5x+2\cdot y \\ &= 28x+21y+10x+2y \\ &= (28x+10x)+(21y+2y) \\ &= 38x+23y \end{aligned}$$

5b. Simplify: $7(3a+2b)+5(4a+2b)$

Objective 1.R.6: Factor trinomials of the form x^2+bx+c.

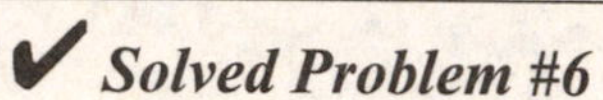

6a. Factor: x^2+5x+6

Factors of 6	6,1	−6,−1	2,3	−2,−3
Sum of Factors	7	−7	5	−5

The factors of 6 whose sum is 5, are 2 and 3.

Thus, $x^2+5x+6=(x+2)(x+3)$.

Check:

$$\begin{aligned} (x+2)(x+3) &= x^2+3x+2x+6 \\ &= x^2+5x+6 \end{aligned}$$

Pencil Problem #6

6a. Factor: $x^2+7x+10$

6b. Factor: $x^2 - 6x + 8$

Factors of 8	8,1	−8,−1	2,4	−2,−4
Sum of Factors	9	−9	6	−6

The factors of 8 whose sum is –6, are –2 and –4.

Thus, $x^2 - 6x + 8 = (x-2)(x-4)$.

Check:

$$\begin{aligned}(x-2)(x-4) &= x^2 - 4x - 2x + 8 \\ &= x^2 - 6x + 8\end{aligned}$$

6b. Factor: $x^2 - 7x + 12$

6c. Factor: $x^2 + 3x - 10$

Factors of −10	−10,1	10,−1	−5,2	5,−2
Sum of Factors	−9	9	−3	3

The factors of –10 whose sum is 3, are 5 and –2.

Thus, $x^2 + 3x - 10 = (x+5)(x-2)$.

Check:

$$\begin{aligned}(x+5)(x-2) &= x^2 - 2x + 5x - 10 \\ &= x^2 + 3x - 10\end{aligned}$$

6c. Factor: $y^2 + 10y - 39$

6d. Factor: $y^2 - 6y - 27$

The factors of –27 whose sum is –6, are –9 and 3.

Thus, $y^2 - 6y - 27 = (y-9)(y+3)$.

6d. Factor: $x^2 - 2x - 15$

6e. Factor: x^2+x-7

No factor pair of –7 has a sum of 1.

Thus, x^2+x-7 is prime.

6e. Factor: $x^2+4x+12$

6f. Factor: $x^2-4xy+3y^2$

The factors of 3 whose sum is –4, are –3 and –1.

Thus, $x^2-4xy+3y^2=(x-3y)(x-y)$.

6f. Factor: $x^2+7xy+6y^2$

6g. Factor: $2x^3+6x^2-56x$

First factor out the common factor of $2x$.

$2x^3+6x^2-56x=2x(x^2+3x-28)$

Continue by factoring the trinomial.

$$\begin{aligned}2x^3+6x^2-56x&=2x(x^2+3x-28)\\&=2x(x-4)(x+7)\end{aligned}$$

6g. Factor: $3x^2+15x+18$

6h. Factor: $-2y^2-10y+28$

First factor out the common factor of -2.

$-2y^2-10y+28=-2(y^2+5y-14)$

Continue by factoring the trinomial.

$$\begin{aligned}-2y^2-10y+28&=-2(y^2+5y-14)\\&=-2(y-2)(y+7)\end{aligned}$$

6h. Factor: $-2x^3-6x^2+8x$

Objective 1.R.7: Factor the difference of two squares.

✔ Solved Problem #7

7a. Factor: $x^2 - 81$

Notice that the trinomial fits the form $A^2 - B^2$.

Thus, factor using $A^2 - B^2 = (A+B)(A-B)$.

$$\begin{aligned} x^2 - 81 &= x^2 - 9^2 \\ &= (x+9)(x-9) \end{aligned}$$

7b. Factor: $36x^2 - 25$

Notice that the trinomial fits the form $A^2 - B^2$.

Thus, factor using $A^2 - B^2 = (A+B)(A-B)$.

$$\begin{aligned} 36x^2 - 25 &= (6x)^2 - 5^2 \\ &= (6x+5)(6x-5) \end{aligned}$$

7c. Factor: $25 - 4x^{10}$

Notice that the trinomial fits the form $A^2 - B^2$.

Thus, factor using $A^2 - B^2 = (A+B)(A-B)$.

$$\begin{aligned} 25 - 4x^{10} &= 5^2 - (2x^5)^2 \\ &= (5+2x^5)(5-2x^5) \end{aligned}$$

Pencil Problem #7

7a. Factor: $x^2 - 25$

7b. Factor: $x^2 + 36$

7c. Factor: $49y^4 - 16$

7d. Factor: $18x^3 - 2x$

First factor out the GCF.

$18x^3 - 2x = 2x(9x^2 - 1)$

Next, factor the difference of two squares.

$$\begin{aligned} 18x^3 - 2x &= 2x(9x^2 - 1) \\ &= 2x(3x+1)(3x-1) \end{aligned}$$

7d. Factor: $2x^3 - 72x$

7e. Factor: $81x^4 - 16$

First, factor the difference of two squares.

$81x^4 - 16 = (9x^2 + 4)(9x^2 - 4)$

The factor of $9x^2 - 4$ is the difference of two squares and can be factored.

$$\begin{aligned} 81x^4 - 16 &= (9x^2 + 4)(9x^2 - 4) \\ &= (9x^2 + 4)(3x+2)(3x-2) \end{aligned}$$

7e. Factor: $x^4 - 16$

Objective 1.R.8: Find numbers for which a rational expression is undefined.

✔ ***Solved Problem #8***

8a. Find all the numbers for which the rational expression is undefined:

$$\frac{7x-28}{8x-40}$$

Set the denominator equal to 0 and solve for x.

$$\begin{aligned} 8x - 40 &= 0 \\ 8x &= 40 \\ x &= 5 \end{aligned}$$

The rational expression is undefined for $x = 5$.

✎ ***Pencil Problem #8*** ✎

8a. Find all the numbers for which the rational expression is undefined:

$$\frac{13}{5x-20}$$

8b. Find all the numbers for which the rational expression is undefined:

$$\frac{8x-40}{x^2+3x-28}$$

Set the denominator equal to 0 and solve for x.

$$x^2+3x-28=0$$
$$(x+7)(x-4)=0$$

$x+7=0$ or $x-4=0$

$x=-7$ $\quad$ $x=4$

The rational expression is undefined for $x=-7$ and $x=4$.

8b. Find all the numbers for which the rational expression is undefined:

$$\frac{y+5}{y^2-25}$$

Objective 1.R.9: Multiply rational expressions.

✔ *Solved Problem #9*

9a. Multiply: $\frac{9}{x+4}\cdot\frac{x-5}{2}$

$$\frac{9}{x+4}\cdot\frac{x-5}{2}=\frac{9(x-5)}{(x+4)2}$$
$$=\frac{9x-45}{2x+8}$$

Pencil Problem #9

9a. Multiply: $\frac{4}{x+3}\cdot\frac{x-5}{9}$

9b. Multiply: $\frac{x+4}{x-7}\cdot\frac{3x-21}{8x+32}$

$$\frac{x+4}{x-7}\cdot\frac{3x-21}{8x+32}=\frac{x+4}{x-7}\cdot\frac{3(x-7)}{8(x+4)}$$

$$=\frac{\overset{1}{\cancel{x+4}}}{\underset{1}{\cancel{x-7}}}\cdot\frac{3\,\overset{1}{\cancel{(x-7)}}}{8\,\underset{1}{\cancel{(x+4)}}}$$

$$=\frac{3}{8}$$

9b. Multiply: $\frac{x-3}{x+5}\cdot\frac{4x+20}{9x-27}$

9c. Multiply: $\frac{x-5}{x-2}\cdot\frac{x^2-4}{9x-45}$

$$\frac{x-5}{x-2}\cdot\frac{x^2-4}{9x-45}=\frac{x-5}{x-2}\cdot\frac{(x+2)(x-2)}{9(x-5)}$$

$$=\frac{\overset{1}{\cancel{x-5}}}{\underset{1}{\cancel{x-2}}}\cdot\frac{(x+2)\,\overset{1}{\cancel{(x-2)}}}{9\,\underset{1}{\cancel{(x-5)}}}$$

$$=\frac{x+2}{9}$$

9c. Multiply: $\frac{x^2-25}{x^2-3x-10}\cdot\frac{x+2}{x}$

9d. Multiply: $\frac{5x+5}{7x-7x^2}\cdot\frac{2x^2+x-3}{4x^2-9}$

$$\frac{5x+5}{7x-7x^2}\cdot\frac{2x^2+x-3}{4x^2-9}=\frac{5(x+1)}{7x(1-x)}\cdot\frac{(2x+3)(x-1)}{(2x+3)(2x-3)}$$

$$=\frac{5(x+1)}{7x\cancel{(1-x)}}\cdot\frac{\cancel{(2x+3)}\,\overset{-1}{\cancel{(x-1)}}}{\cancel{(2x+3)}(2x-3)}$$

$$=\frac{-5(x+1)}{7x(2x-3)}\text{ or }-\frac{5(x+1)}{7x(2x-3)}$$

9d. Multiply: $\frac{25-y^2}{y^2-2y-35}\cdot\frac{y^2-8y-20}{y^2-3y-10}$

Objective 1.R.10: Find the least common denominator.

✔ Solved Problem #10

10a. Find the LCD of $\frac{3}{10x^2}$ and $\frac{7}{15x}$.

List the factors for each denominator.

$10x^2 = 2 \cdot 5x^2$

$15x = 3 \cdot 5x$

LCD = $2 \cdot 3 \cdot 5 \cdot x^2 = 30x^2$

10b. Find the LCD of $\frac{2}{x+3}$ and $\frac{4}{x-3}$.

List the factors for each denominator.

$x+3 = 1(x+3)$

$x-3 = 1(x-3)$

LCD = $(x+3)(x-3)$

10c. Find the LCD of $\frac{9}{7x^2+28x}$ and $\frac{11}{x^2+8x+16}$.

List the factors for each denominator.

$7x^2+28x = 7x(x+4)$

$x^2+8x+16 = (x+4)^2$

LCD = $7x(x+4)^2$

Pencil Problem #10

10a. Find the LCD of $\frac{8}{15x^2}$ and $\frac{5}{6x^5}$.

10b. Find the LCD of $\frac{4}{x-3}$ and $\frac{7}{x+1}$.

10c. Find the LCD of $\frac{7}{y^2-1}$ and $\frac{y}{y^2-2y+1}$.

Objective 1.R.11: Translate English phrases into algebraic expressions.

✔ *Solved Problem #11*

11. Write each English phrase as an algebraic expression. Let the variable x represent the number.

11a. The product of 6 and a number

$6x$

11b. A number added to 4

$4+x$

11c. Three times a number, increased by 5

$3x+5$

11d. Twice a number subtracted from 12

$12-2x$

11e. The quotient of 15 and a number

$\frac{15}{x}$

Pencil Problem #11

11. Write each English phrase as an algebraic expression. Let the variable x represent the number.

11a. Four more than a number

11b. Nine subtracted from a number

11c. Three times a number, decreased by 5

11d. One less than the product of 12 and a number

11e. Six more than the quotient of a number and 30

Objective 1.R.12: Translate English sentences into algebraic equations.

✔ *Solved Problem #12*

12a. Write the sentence as an equation. Let the variable x represent the number.

The quotient of a number and 6 is 5.

$$\frac{x}{6}=5$$

12b. Write the sentence as an equation. Let the variable x represent the number.

Seven decreased by twice a number yields 1.

$$7-2x=1$$

Pencil Problem #12

12a. Write the sentence as an equation. Let the variable x represent the number.

Four times a number is 28.

12b. Write the sentence as an equation. Let the variable x represent the number.

Five times a number is equal to 24 decreased by the number.

Objective 1.R.13: Use the percent formula.

13a. What is 9% of 50?

Use the formula $A=PB$: A is P percent of B.

What	is	9%	of	50?
A	$=$	0.09	$\cdot$	50

$$A=4.5$$

4.5 is 9% of 50.

13b. 9 is 60% of what?

Use the formula $A=PB$: A is P percent of B.

9	is	60%	of	what?
9	$=$	0.60	$\cdot$	B

$$\frac{9}{0.60}=\frac{0.60B}{0.60}$$

$$15=B$$

9 is 60% of 15.

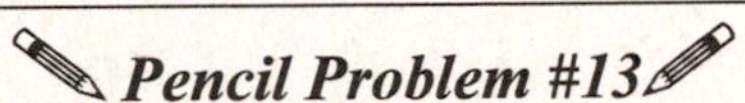

13a. What is 3% of 200?

13b. 24% of what number is 40.8?

13c. 18 is what percent of 50?

Use the formula $A = PB$: A is P percent of B.

$$\underbrace{\boxed{18}}_{18}\ \underbrace{\boxed{\text{is}}}_{=}\ \underbrace{\boxed{\text{what percent}}}_{P}\ \underbrace{\boxed{\text{of}}}_{\cdot}\ \underbrace{\boxed{50?}}_{50}$$

$18 = P \cdot 50$

$\frac{18}{50} = \frac{50P}{50}$

$0.36 = P$

To change 0.36 to a percent, move the decimal point two places to the right and add a percent sign.

$0.36 = 36\%$

18 is 36% of 50.

13c. 3 is what percent of 15?

Objective 1.R.14: Multiply radical expressions with more than one term.

✔ *Solved Problem #14*

14a. Multiply: $\sqrt{2}\left(\sqrt{5}+\sqrt{11}\right)$

Use the distributive property.

$$\begin{aligned}\sqrt{2}\left(\sqrt{5}+\sqrt{11}\right) &= \sqrt{2}\sqrt{5}+\sqrt{2}\sqrt{11}\\ &= \sqrt{10}+\sqrt{22}\end{aligned}$$

14b. Multiply: $(4+\sqrt{3})(2+\sqrt{3})$

Use FOIL.

$$\begin{aligned}(4+\sqrt{3})(2+\sqrt{3}) &= 4\cdot 2+4\sqrt{3}+2\sqrt{3}+\left(\sqrt{3}\right)^2\\ &= 8+4\sqrt{3}+2\sqrt{3}+3\\ &= 11+6\sqrt{3}\end{aligned}$$

Pencil Problem #14

14a. Multiply: $\sqrt{7}\left(\sqrt{6}-\sqrt{10}\right)$

14b. Multiply: $(5+\sqrt{2})(6+\sqrt{2})$

14c. Multiply: $(3+\sqrt{5})(8-4\sqrt{5})$

Use FOIL.

$$\begin{aligned}(3+\sqrt{5})(8-4\sqrt{5}) &= 3\cdot 8-3\cdot 4\sqrt{5}+8\sqrt{5}-4\left(\sqrt{5}\right)^2\\ &= 24-12\sqrt{5}+8\sqrt{5}-4\cdot 5\\ &= 24-12\sqrt{5}+8\sqrt{5}-20\\ &= 4-4\sqrt{5}\end{aligned}$$

14c. Multiply: $(6-3\sqrt{7})(2-5\sqrt{7})$

Objective 1.R.15: Multiply conjugates.

✔ ***Solved Problem #15***

15a. Multiply: $(3+\sqrt{11})(3-\sqrt{11})$

Use the special-product formula
$(A+B)(A-B)=A^2-B^2$.

$$\begin{aligned}(3+\sqrt{11})(3-\sqrt{11}) &= 3^2-\left(\sqrt{11}\right)^2\\ &= 9-11\\ &= -2\end{aligned}$$

15b. Multiply: $(\sqrt{7}-\sqrt{2})(\sqrt{7}+\sqrt{2})$

Use the special-product formula
$(A+B)(A-B)=A^2-B^2$.

$$\begin{aligned}(\sqrt{7}-\sqrt{2})(\sqrt{7}+\sqrt{2}) &= \left(\sqrt{7}\right)^2-\left(\sqrt{2}\right)^2\\ &= 7-2\\ &= 5\end{aligned}$$

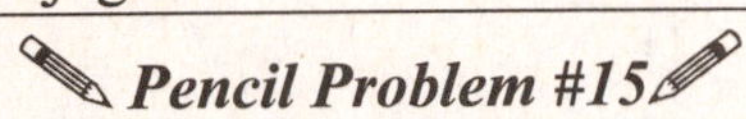

15a. Multiply: $(1-\sqrt{6})(1+\sqrt{6})$

15b. Multiply: $(2\sqrt{3}+7)(2\sqrt{3}-7)$

Objective 1.R.16: Rationalize denominators containing two terms.

✔ Solved Problem #16

16a. Rationalize the denominator: $\frac{8}{4+\sqrt{5}}$

Multiply the numerator and the denominator by the conjugate of the denominator.

$$\frac{8}{4+\sqrt{5}} = \frac{8}{4+\sqrt{5}} \cdot \frac{4-\sqrt{5}}{4-\sqrt{5}}$$
$$= \frac{32-8\sqrt{5}}{16-5}$$
$$= \frac{32-8\sqrt{5}}{11}$$

16b. Rationalize the denominator: $\frac{8}{\sqrt{7}-\sqrt{3}}$

Multiply the numerator and the denominator by the conjugate of the denominator.

$$\frac{8}{\sqrt{7}-\sqrt{3}} = \frac{8}{\sqrt{7}-\sqrt{3}} \cdot \frac{\sqrt{7}+\sqrt{3}}{\sqrt{7}+\sqrt{3}}$$
$$= \frac{8\sqrt{7}+8\sqrt{3}}{(\sqrt{7})^2-(\sqrt{3})^2}$$
$$= \frac{8\sqrt{7}+8\sqrt{3}}{7-3}$$
$$= \frac{4(2\sqrt{7}+2\sqrt{3})}{4}$$
$$= 2\sqrt{7}+2\sqrt{3}$$

Pencil Problem #16

16a. Rationalize the denominator: $\frac{1}{4+\sqrt{3}}$

16b. Rationalize the denominator: $\frac{6}{\sqrt{6}+\sqrt{3}}$

Objective 1.R.17: Find the greatest common factor.

✔ *Solved Problem #17*

17a. Find the greatest common factor of the following list of monomials: $18x^3$ and $15x^2$

$18x^3 = 3x^2 \cdot 6x$

$15x^2 = 3x^2 \cdot 5$

The GCF is $3x^2$.

17b. Find the greatest common factor of the following list of monomials: x^4y, x^3y^2, and x^2y

$x^4y = x^2y \cdot x^2$

$x^3y^2 = x^2y \cdot xy$

$x^2y = x^2y$

The GCF is x^2y.

✎ *Pencil Problem #17* ✎

17a. Find the greatest common factor of the following list of monomials: $12x^2$ and $8x$

17b. Find the greatest common factor of the following list of monomials: $16x^5y^4$, $8x^6y^3$, and $20x^4y^5$

Objective 1.R.18: Factor out the greatest common factor of a polynomial.

✔ ***Solved Problem #18***	✎ ***Pencil Problem #18***
18a. Factor: $6x^2+18$ The GCF is 6. $6x^2+18=6\cdot x^2+6\cdot 3$ $=6(x^2+3)$	**18a.** Factor: $18y^2+12$
18b. Factor: $25x^2+35x^3$ The GCF is $5x^2$. $25x^2+35x^3=5x^2\cdot 5+5x^2\cdot 7x$ $=5x^2(5+7x)$	**18b.** Factor: $8x^2-4x^4$
18c. Factor: $15x^5+12x^4-27x^3$ The GCF is $3x^3$. $15x^5+12x^4-27x^3=3x^3\cdot 5x^2+3x^3\cdot 4x-3x^3\cdot 9$ $=3x^3(5x^2+4x-9)$	**18c.** Factor: $100y^5-50y^3+100y^2$
18d. Factor: $8x^3y^2-14x^2y+2xy$ The GCF is $2xy$. $8x^3y^2-14x^2y+2xy=2xy\cdot 4x^2y-2xy\cdot 7x+2xy\cdot 1$ $=2xy(4x^2y-7x+1)$	**18d.** Factor: $11x^2-23$

Objective R.1.19: Factor trinomials of the form $ax^2 + bx + c$.

✔ *Solved Problem #19*

19a. Factor $6x^2 + 19x - 7$ by trial and error.

Step 1 Find two first terms whose product is $6x^2$.

$$6x^2 + 19x - 7 = (6x \quad)(x \quad)$$

$$6x^2 + 19x - 7 = (3x \quad)(2x \quad)$$

Step 2 The last term, –7, has possible factorizations of $1(-7)$ and $-1(7)$.

Step 3

Possible Factors of $6x^2 + 19x - 7$	Sum of Outside and Inside Products
$(6x+1)(x-7)$	$-42x + x = -41x$
$(6x-7)(x+1)$	$6x - 7x = -x$
$(6x-1)(x+7)$	$42x - x = 41x$
$(6x+7)(x-1)$	$-6x + 7x = x$
$(3x+1)(2x-7)$	$-21x + 2x = -19x$
$(3x-7)(2x+1)$	$3x - 14x = -11x$
$(3x-1)(2x+7)$	required middle term $\overbrace{21x - 2x = 19x}$
$(3x+7)(2x-1)$	$-3x + 14x = 11x$

The required middle term is obtained by using the factors $(3x-1)(2x+7)$.

Check: $(3x-1)(2x+7) = 6x^2 + 21x - 2x - 7$

$$= 6x^2 + 19x - 7$$

Thus, $6x^2 + 19x - 7 = (3x-1)(2x+7)$

Pencil Problem #19

19a. Factor $2x^2 + 5x + 3$ by trial and error.

19b. Factor $3x^2 - 13xy + 4y^2$ by trial and error.

Step 1 Find two First terms whose product is $3x^2$.

$$3x^2 - 13xy + 4y^2 = (3x \quad)(x \quad)$$

Step 2 The last term, $4y^2$, has pairs of factors that are either both positive or both negative. Because the middle term, $-13xy$, is negative, both factors must be negative. Thus the last term has possible factorizations of $-2y(-2y)$ or $-y(-4y)$.

Step 3

Possible Factors of $3x^2 - 13xy + 4y^2$	Sum of Outside and Inside Products
$(3x-4y)(x-y)$	$-3xy - 4xy = -7xy$
$(3x-y)(x-4y)$	$\overbrace{-12xy - xy = -13xy}^{\text{required middle term}}$
$(3x-2y)(x-2y)$	$-6xy - 2xy = -8xy$

The required middle term is obtained by using the factors $(3x-y)(x-4y)$.

Check: $(3x-y)(x-4y) = 3x^2 - 12xy - xy + 4y^2$

$$= 3x^2 - 13xy + 4y^2$$

Thus, $3x^2 - 13xy + 4y^2 = (3x-y)(x-4y)$

19b. Factor $3x^2 + 5xy + 2y^2$ by trial and error.

Objective 1.R.20: Factor perfect square trinomials.

Solved Problem #20

20a. Factor: $x^2 + 14x + 49$

Notice that the trinomial fits the form $A^2 + 2AB + B^2$.

Thus, factor using $A^2 + 2AB + B^2 = (A + B)^2$.

$x^2 + 14x + 49 = (x + 7)^2$

20b. Factor: $16x^2 - 56x + 49$

Notice that the trinomial fits the form $A^2 - 2AB + B^2$.

Thus, factor using $A^2 - 2AB + B^2 = (A - B)^2$.

$16x^2 - 56x + 49 = (4x - 7)^2$

20c. Factor: $4x^2 + 12xy + 9y^2$

Notice that the trinomial fits the form $A^2 + 2AB + B^2$.

Thus, factor using $A^2 + 2AB + B^2 = (A + B)^2$.

$4x^2 + 12xy + 9y^2 = (2x + 3y)^2$

Pencil Problem #20

20a. Factor: $x^2 + 2x + 1$

20b. Factor: $25y^2 - 10y + 1$

20c. Factor: $16x^2 - 40xy + 25y^2$

Objective R.1.21: Simplify square roots.

✔ *Solved Problem #21*

21a. Simplify: $\sqrt{60}$

4 is the greatest perfect square that is a factor of 60.

$$\begin{aligned}\sqrt{60} &= \sqrt{4\cdot 15}\\ &= \sqrt{4}\sqrt{15}\\ &= 2\sqrt{15}\end{aligned}$$

21b. Simplify: $\sqrt{55}$

$\sqrt{55}$ cannot be simplified because it has no perfect square factors other than 1.

21c. Simplify: $\sqrt{40x^{16}}$

$4x^{16}$ is the greatest perfect square that is a factor of $40x^{16}$.

$$\begin{aligned}\sqrt{40x^{16}} &= \sqrt{4x^{16}\cdot 10}\\ &= 2x^8\sqrt{10}\end{aligned}$$

21d. Multiply and then simplify: $\sqrt{15x^6}\cdot\sqrt{3x^7}$

$$\begin{aligned}\sqrt{15x^6}\cdot\sqrt{3x^7} &= \sqrt{15x^6\cdot 3x^7}\\ &= \sqrt{45x^{13}}\\ &= \sqrt{9x^{12}\cdot 5x}\\ &= 3x^6\sqrt{5x}\end{aligned}$$

Pencil Problem #21

21a. Simplify: $\sqrt{50}$

21b. Simplify: $\sqrt{35}$

21c. Simplify: $\sqrt{20x^6}$

21d. Multiply and then simplify: $\sqrt{15x^4}\cdot\sqrt{5x^9}$

Objective 1.R.22: Find the square of a binomial sum.

✔ Solved Problem #22

22a. Multiply: $(x+10)^2$

Use the special-product formula $(A+B)^2 = A^2 + 2AB + B^2$.

$$(x+10)^2 = \overbrace{x^2}^{\text{first term squared}} + \overbrace{2 \cdot 10x}^{2\cdot\text{product of the terms}} + \overbrace{10^2}^{\text{last term squared}}$$
$$= x^2 + 20x + 100$$

22b. Multiply: $(5x+4)^2$

Use the special-product formula $(A+B)^2 = A^2 + 2AB + B^2$.

$$(5x+4)^2 = \overbrace{(5x)^2}^{\text{first term squared}} + \overbrace{2 \cdot 20x}^{2\cdot\text{product of the terms}} + \overbrace{4^2}^{\text{last term squared}}$$
$$= 25x^2 + 40x + 16$$

Pencil Problem #22

22a. Multiply: $(x+2)^2$

22b. Multiply: $(x^8+3)^2$

Objective 1.R.23: Find the square of a binomial difference.

✔ *Solved Problem #23*

23a. Multiply: $(x-9)^2$

Use the special-product formula
$(A-B)^2 = A^2 - 2AB + B^2$.

$$(x-9)^2 = \overbrace{x^2}^{\text{first term squared}} - \overbrace{2\cdot 9x}^{2\cdot\text{product of the terms}} + \overbrace{9^2}^{\text{last term squared}}$$
$$= x^2 - 18x + 81$$

23b. Multiply: $(7x-3)^2$

Use the special-product formula
$(A-B)^2 = A^2 - 2AB + B^2$.

$$(7x-3)^2 = \overbrace{(7x)^2}^{\text{first term squared}} - \overbrace{2\cdot 21x}^{2\cdot\text{product of the terms}} + \overbrace{3^2}^{\text{last term squared}}$$
$$= 49x^2 - 42x + 9$$

Pencil Problem #23

23a. Multiply: $(x-3)^2$

23b. Multiply: $(7-2x)^2$

Objective 1.R.24: Factor by grouping.

✔ *Solved Problem #24*

24a. Factor: $x^2(x+1)+7(x+1)$

Factor out the greatest common factor of $x+1$.

$$x^2\overbrace{(x+1)}^{\text{GCF}} + 7\overbrace{(x+1)}^{\text{GCF}} = (x+1)(x^2+7)$$

24b. Factor: $x(y+4)-7(y+4)$

Factor out the greatest common factor of $y+4$.

$$x\overbrace{(y+4)}^{\text{GCF}} - 7\overbrace{(y+4)}^{\text{GCF}} = (y+4)(x-7)$$

Pencil Problem #24

24a. Factor: $x(x+5)+3(x+5)$

24b. Factor: $3x(x+y)-(x+y)$

24c. Factor: $x^3+5x^2+2x+10$

There is no factor other than 1 common to all four terms. However, we can group terms that have a common factor.

$x^3+5x^2+2x+10=(x^3+5x^2)+(2x+10)$

Factor out the greatest common factor from the grouped terms. The remaining two terms have $x+5$ as a common binomial factor, which should then be factored out.

$$\begin{aligned} x^3+5x^2+2x+10 &= (x^3+5x^2)+(2x+10) \\ &= x^2(x+5)+2(x+5) \\ &= (x+5)(x^2+2) \end{aligned}$$

24c. Factor: $x^2+2x+4x+8$

24d. Factor: $xy+3x-5y-15$

There is no factor other than 1 common to all four terms. However, we can group terms that have a common factor.

$xy+3x-5y-15=(xy+3x)+(-5y-15)$

Factor out x from the first two grouped terms and -5 from the last two grouped terms.

$xy+3x-5y-15=x(y+3)-5(y+3)$

Finally, factor out the common factor of $y+3$.

$$\begin{aligned} xy+3x-5y-15 &= x(y+3)-5(y+3) \\ &= (y+3)(x-5) \end{aligned}$$

24d. Factor: $3x^3-2x^2-6x+4$

Objective 1.R.25: Evaluate expressions with rational exponents.

✔ *Solved Problem #25*

25a. Simplify: $25^{\frac{1}{2}}$

The denominator of the exponent is the index of the radical. Thus an exponent of $\frac{1}{2}$ is equivalent to a square root.

$$\begin{aligned} 25^{\frac{1}{2}} &= \sqrt{25} \\ &= 5 \end{aligned}$$

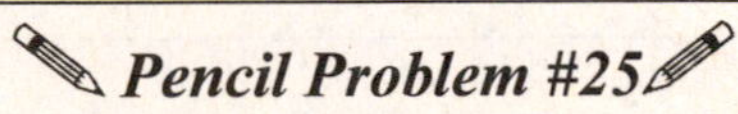

Pencil Problem #25

25a. Simplify: $49^{\frac{1}{2}}$

25b. Simplify: $-81^{\frac{1}{4}}$

The denominator of the exponent is the index of the radical.

Note that the base is 81 and the negative sign is not affected by the exponent.

$$-81^{\frac{1}{4}} = -\sqrt[4]{81}$$
$$= -3$$

25b. Simplify: $-125^{\frac{1}{3}}$

25c. Simplify: $(-8)^{\frac{1}{3}}$

The denominator of the exponent is the index of the radical.

Note that the parentheses indicated that the base is -8. Thus the negative sign is affected by the exponent and should by under the radical.

$$(-8)^{\frac{1}{3}} = \sqrt[3]{-8}$$
$$= -2$$

25c. Simplify: $\left(\frac{27}{64}\right)^{\frac{1}{3}}$

25d. Simplify: $27^{\frac{4}{3}}$

The denominator, 3, is the radical's index and the numerator, 4, is the exponent.

$$27^{\frac{4}{3}} = \left(\sqrt[3]{27}\right)^4$$
$$= 3^4$$
$$= 81$$

25d. Simplify: $81^{\frac{3}{2}}$

25e. Simplify: $4^{\frac{3}{2}}$

The denominator, 2, is the radical's index and the numerator, 3, is the exponent.

$$4^{\frac{3}{2}} = \left(\sqrt{4}\right)^3 = 2^3 = 8$$

25e. Simplify: $9^{\frac{3}{2}}$

25f. Simplify: $25^{-\frac{1}{2}}$

$$25^{-\frac{1}{2}} = \frac{1}{25^{\frac{1}{2}}} = \frac{1}{\sqrt{25}} = \frac{1}{5}$$

25f. Simplify: $9^{-\frac{1}{2}}$

25g. Simplify: $32^{-\frac{4}{5}}$

$$32^{-\frac{4}{5}} = \frac{1}{32^{\frac{4}{5}}} = \frac{1}{\left(\sqrt[5]{32}\right)^4} = \frac{1}{2^4} = \frac{1}{16}$$

25g. Simplify: $81^{-\frac{5}{4}}$

Objective 1.R.26: Rewrite radical expressions with rational exponents.

✔ Solved Problem #26

26a. Rewrite using radical notation and simplify: $8^{\frac{4}{3}}$

$$8^{\frac{4}{3}} = \left(\sqrt[3]{8}\right)^4$$
$$= 2^4$$
$$= 16$$

26b. Rewrite using rational exponents: $\left(\sqrt[5]{2xy}\right)^7$

$$\left(\sqrt[5]{2xy}\right)^7 = (2xy)^{\frac{7}{5}}$$

Pencil Problem #26

26a. Rewrite using radical notation and simplify: $81^{\frac{3}{2}}$

26b. Rewrite using rational exponents: $2x\sqrt[3]{y^2}$

Objective 1.R.27: Graph numbers on a number line.

✔ Solved Problem #27

27a. Graph: –4

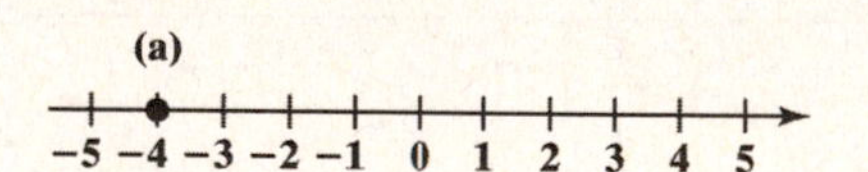

27b. Graph: –1.2

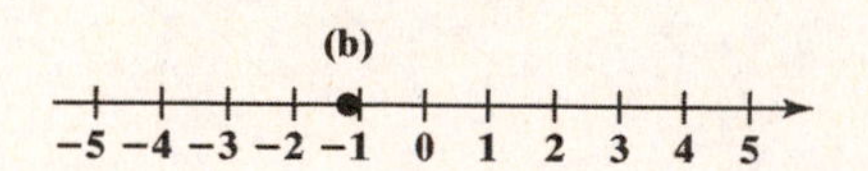

Pencil Problem #27

27a. Graph: 2

27b. Graph: $-\frac{16}{5}$

Objective 1.R.28: Understand and use inequality symbols.

✔ *Solved Problem #28*

28a. Insert either $<$ or $>$ to make the statement true.
$-19 \quad -6$

Since -19 is to the left of -6 on the number line, then $-19<-6$.

28b. Determine if the inequality is true or false.
$-2 \geq -2$

Because $-2=-2$ is true, then $-2 \geq -2$ is true.

28c. Determine if the inequality is true or false.
$-4 \geq 1$

Because neither $-4>1$ nor $-4=1$ is true, then $-4 \geq 1$ is false.

Pencil Problem #28

28a. Insert either $<$ or $>$ to make the statement true.
$-\pi \quad -3.5$

28b. Determine if the inequality is true or false.
$0 \geq -6$

28c. Determine if the inequality is true or false.
$-17 \geq 6$

Answers for Pencil Problems:

1a. 17 **1b.** 5

2. $A\ (5,2)$, $C\ (-6,5)$, $E(-2,-3)$

3.

x	$y=-3x+7$	(x,y)
-2	$y=-3(-2)+7=13$	$(-2,13)$
-1	$y=-3(-1)+7=10$	$(-1,10)$
0	$y=-3(0)+7=7$	$(0,7)$
1	$y=-3(1)+7=4$	$(1,4)$
2	$y=-3(2)+7=1$	$(2,1)$

4a. Solution **4b.** Not a solution

5a. $15x+6$ **5b.** $41a+24b$

6a. $(x+5)(x+2)$ **6b.** $(x-4)(x-3)$ **6c.** $(y+13)(y-3)$ **6d.** $(x-5)(x+3)$

6e. Prime **6f.** $(x+6y)(x+y)$ **6g.** $3(x+2)(x+3)$ **6h.** $-2x(x+4)(x-1)$

7a. $(x+5)(x-5)$ **7b.** Prime **7c.** $(7y^2+4)(7y^2-4)$

7d. $2x(x+6)(x-6)$ **7e.** $(x^2+4)(x+2)(x-2)$

8a. $x=4$ **8b.** $y=-5$ and $y=5$

9a. $\frac{4x-20}{9x+27}$ **9b.** $\frac{4}{9}$ **9c.** $\frac{x+5}{x}$ **9d.** $\frac{-(y-10)}{y-7}$ or $-\frac{y-10}{y-7}$

10a. $2\cdot3\cdot5\cdot x^5=30x^5$ **10b.** $(x-3)(x+1)$ **10c.** $(y+1)(y-1)(y-1)$

11a. $x+4$ **11b.** $x-9$ **11c.** $3x-5$ **11d.** $12x-1$ **11e.** $\frac{x}{30}+6$

12a. $4x=28$ **12b.** $5x=24-x$

13a. 6 **13b.** 170 **13c.** 20%

14a. $\sqrt{42}-\sqrt{70}$ **14b.** $32+11\sqrt{2}$ **14c.** $117-36\sqrt{7}$

15a. -5 **15b.** -37

16a. $\frac{4-\sqrt{3}}{13}$ **16b.** $2\sqrt{6}-2\sqrt{3}$

17a. $4x$ **17b.** $4x^4y^3$

18a. $6(3y^2+2)$ **18b.** $4x^2(2-x^2)$ **18c.** $50y^2(2y^3-y+2)$ **18d.** Cannot be factored

19a. $(2x+3)(x+1)$ **19b.** $(3x+2y)(x+y)$

20a. $(x+1)^2$ **20b.** $(5y-1)^2$ **20c.** $(4x-5y)^2$

21a. $5\sqrt{2}$ **21b.** $\sqrt{35}$ Cannot be simplified **21c.** $2x^3\sqrt{5}$ **21d.** $5x^6\sqrt{3x}$

22a. x^2+4x+4 **22b.** $x^{16}+6x^8+9$

23a. x^2-6x+9 **23b.** $49-28x+4x^2$

24a. $(x+5)(x+3)$ **24b.** $(x+y)(3x-1)$ **24c.** $(x+2)(x+4)$ **24d.** $(3x-2)(x^2-2)$

25a. 7 **25b.** -5 **25c.** $\frac{3}{4}$ **25d.** 729 **25e.** 27 **25f.** $\frac{1}{3}$

25g. $\frac{1}{243}$

26a. 729 **26b.** $2xy^{2/3}$

27a. −5 −4 −3 −2 −1 0 1 2 3 4 5 **27b.** −5 −4 −3 −2 −1 0 1 2 3 4 5

28a. $-\pi>-3.5$ **28b.** True **28c.** False

Chapter 2
Functions and Graphs

Guided Practice:

☐ Review each of the following ***Solved Problems*** and complete each ***Pencil Problem***.

Objective 2.R.1: Use the order of operations.

✔ *Solved Problem #1*

1. Simplify: $3-5^2+12\div 2(-4)^2$

$$\begin{aligned}3-5^2+12\div 2(-4)^2 &= 3-25+12\div 2(16)\\ &= 3-25+6(16)\\ &= 3-25+96\\ &= -22+96\\ &= 74\end{aligned}$$

Pencil Problem #1

1. Simplify: $15-\sqrt{3-(-1)}+12\div 2\cdot 3$

Objective 2.R.2: Graph equations in the rectangular coordinate system.

✔ *Solved Problem #2*

2a. Graph $y=1-x^2$.

x	$y=1-x^2$	(x,y)
-3	$y=1-(-3)^2=-8$	$(-3,-8)$
-2	$y=1-(-2)^2=-3$	$(-2,-3)$
-1	$y=1-(-1)^2=0$	$(-1,0)$
0	$y=1-(0)^2=1$	$(0,1)$
1	$y=1-(1)^2=0$	$(1,0)$
2	$y=1-(2)^2=-3$	$(2,-3)$
3	$y=1-(3)^2=-8$	$(3,-8)$

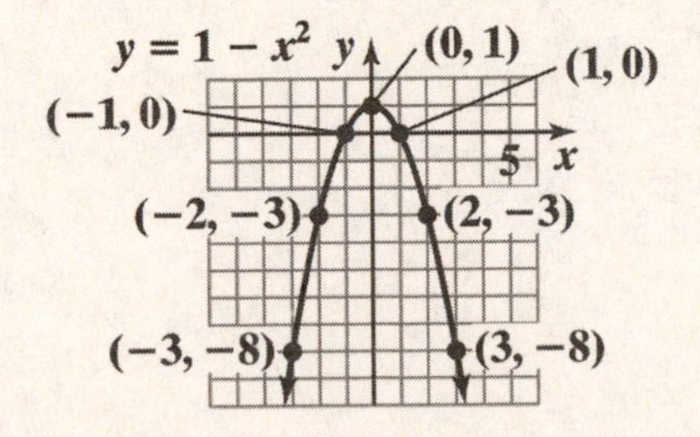

Pencil Problem #2

2a. Graph $y=x^2-4$. Let $x=-3,-2,-1,0,1,2,$ and 3.

2b. Graph $y=|x+1|$.

x	$y=\|x+1\|$	(x,y)
-4	$y=\|-4+1\|=\|-3\|=3$	$(-4,3)$
-3	$y=\|-3+1\|=\|-2\|=2$	$(-3,2)$
-2	$y=\|-2+1\|=\|-1\|=1$	$(-2,1)$
-1	$y=\|-1+1\|=\|0\|=0$	$(-1,0)$
0	$y=\|0+1\|=\|1\|=1$	$(0,1)$
1	$y=\|1+1\|=\|2\|=2$	$(1,2)$
2	$y=\|2+1\|=\|3\|=3$	$(2,3)$

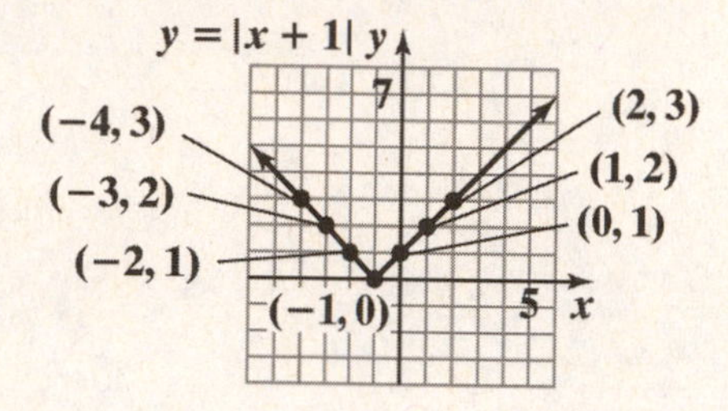

2b. Graph $y=2|x|$. Let $x=-3,-2,-1,0,1,2,$ and 3.

Objective 2.R.3: Solve a formula for a variable.

✔ *Solved Problem #3*

3a. Solve the formula $2l+2w=P$ for w.

$$2l+2w=P$$
$$2w=P-2l$$
$$\frac{2w}{2}=\frac{P-2l}{2}$$
$$w=\frac{P-2l}{2}$$

3b. Solve the formula $P=C+MC$ for C.

$$P=C+MC$$
$$P=C(1+M)$$
$$\frac{P}{1+M}=\frac{C(1+M)}{1+M}$$
$$\frac{P}{1+M}=C$$
$$C=\frac{P}{1+M}$$

Pencil Problem #3

3a. Solve the formula $T=D+pm$ for p.

3b. Solve the formula $IR+Ir=E$ for I.

Objective 2.R.4: Square binomials.

✔ *Solved Problem #4*

4a. Multiply: $(x+8)^2$

Use the special-product formula
$(A+B)^2 = A^2 + 2AB + B^2$.

$$(x+8)^2 = x^2 + 2 \cdot x \cdot 8 + 8^2$$
$$= x^2 + 16x + 64$$

4b. Multiply: $(2x - 6y^4)^2$

Use the special-product formula
$(A-B)^2 = A^2 - 2AB + B^2$.

$$(2x-6y^4)^2 = (2x)^2 - 2 \cdot 2x \cdot 6y^4 + (6y^4)^2$$
$$= 4x^2 - 24xy^4 + 36y^8$$

Pencil Problem #4

4a. Multiply: $(2x+y)^2$

4b. Multiply: $(y-5)^2$

Objective 2.R.5: Simplify rational expressions.

✔ *Solved Problem #5*

5a. Simplify: $\dfrac{x^2+7x+10}{x+2}$

$$\frac{x^2+7x+10}{x+2} = \frac{(x+5)(x+2)}{1(x+2)}$$
$$= \frac{(x+5)\cancel{(x+2)}}{1\cancel{(x+2)}}$$
$$= x+5$$

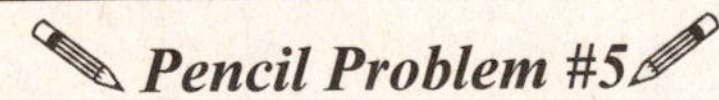

Pencil Problem #5

5a. Simplify: $\dfrac{x+2}{x^2-x-6}$

5b. Simplify: $\dfrac{x^2-2x-15}{3x^2+8x-3}$

$$\frac{x^2-2x-15}{3x^2+8x-3}=\frac{(x-5)(x+3)}{(3x-1)(x+3)}$$
$$=\frac{(x-5)\cancel{(x+3)}}{(3x-1)\cancel{(x+3)}}$$
$$=\frac{x-5}{3x-1}$$

5b. Simplify: $\dfrac{4x+20}{x^2+5x}$

Objective 2.R.6: Use point plotting to graph linear equations.

✔ *Solved Problem #6*

6a. Graph the equation: $y=2x$

First, make a table of values:

x	$y=2x$	(x,y)
-2	$y=2(-2)=-4$	$(-2,-4)$
-1	$y=2(-1)=-2$	$(-1,-2)$
0	$y=2(0)=0$	$(0,0)$
1	$y=2(1)=2$	$(1,2)$
2	$y=2(2)=4$	$(2,4)$

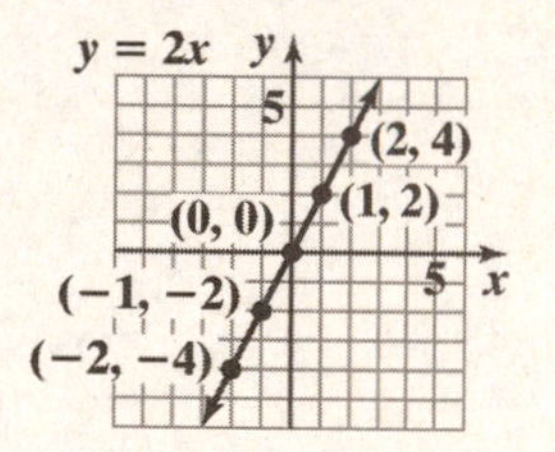

Pencil Problem #6

6a. Graph the equation: $y=x$

6b. Graph the equation: $y=\frac{1}{2}x+2$

First, make a table of values:

x	$y=\frac{1}{2}x+2$	(x,y)
-4	$y=\frac{1}{2}(-4)+2=0$	$(-4,0)$
-2	$y=\frac{1}{2}(-2)+2=1$	$(-2,1)$
0	$y=\frac{1}{2}(0)+2=2$	$(0,2)$
2	$y=\frac{1}{2}(2)+2=3$	$(2,3)$
4	$y=\frac{1}{2}(4)+2=4$	$(4,4)$

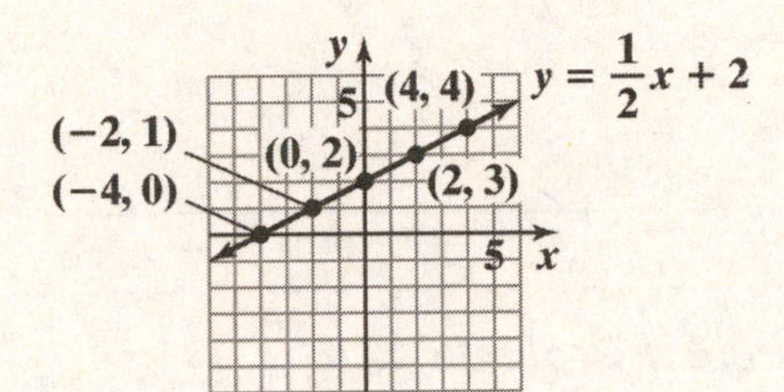

6b. Graph the equation: $y=-\frac{3}{2}x+1$

Objective 2.R.7: Solve a formula for a variable.

Solved Problem #7

7a. Solve the formula $2l+2w=P$ for w.

$$2l+2w=P$$
$$2w=P-2l$$
$$\frac{2w}{2}=\frac{P-2l}{2}$$
$$w=\frac{P-2l}{2}$$

Pencil Problem #7

7a. Solve the formula $T=D+pm$ for p.

7b. Solve the formula $P = C + MC$ for C.

$$P = C + MC$$
$$P = C(1+M)$$
$$\frac{P}{1+M} = \frac{C(1+M)}{1+M}$$
$$\frac{P}{1+M} = C$$
$$C = \frac{P}{1+M}$$

7b. Solve the formula $IR + Ir = E$ for I.

Objective 2.R.8: Add polynomials.

✔ *Solved Problem #8*

8a. Add:
$(-7x^3 + 4x^2 + 3) + (4x^3 + 6x^2 - 13)$

$$(-7x^3 + 4x^2 + 3) + (4x^3 + 6x^2 - 13)$$
$$= -7x^3 + 4x^2 + 3 + 4x^3 + 6x^2 - 13$$
$$= \underbrace{-7x^3 + 4x^3}_{-3x^3} + \underbrace{4x^2 + 6x^2}_{+10x^2} + \underbrace{3 - 13}_{-10}$$
$$= -3x^3 + 10x^2 - 10$$

8b. Add by aligning vertically:
$(7xy^3 - 5xy^2 - 3y) + (2xy^3 + 8xy^2 - 12y - 9)$

$$\begin{array}{l} 7xy^3 - 5xy^2 - 3y \\ \underline{2xy^3 + 8xy^2 - 12y - 9} \\ 9xy^3 + 3xy^2 - 15y - 9 \end{array}$$

Pencil Problem #8

8a. Add:
$(-6x^3 + 5x^2 - 8x + 9) + (17x^3 + 2x^2 - 4x - 13)$

8b. Add by aligning vertically:
$(7x^2y - 5xy) + (2x^2y - xy)$

Objective 2.R.9: Subtract polynomials.

✔ *Solved Problem #9*

9a. Subtract:

$(14x^3 - 5x^2 + x - 9) - (4x^3 - 3x^2 - 7x + 1)$

$$\begin{aligned}
&(14x^3 - 5x^2 + x - 9) - (4x^3 - 3x^2 - 7x + 1) \\
&= (14x^3 - 5x^2 + x - 9) + (-4x^3 + 3x^2 + 7x - 1) \\
&= 14x^3 - 5x^2 + x - 9 - 4x^3 + 3x^2 + 7x - 1 \\
&= 14x^3 - 4x^3 - 5x^2 + 3x^2 + x + 7x - 9 - 1 \\
&= 10x^3 - 2x^2 + 8x - 10
\end{aligned}$$

9b. Subtract $-7x^2y^5 - 4xy^3 + 2$ from $6x^2y^5 - 2xy^3 - 8$.

$$\begin{aligned}
&(6x^2y^5 - 2xy^3 - 8) - (-7x^2y^5 - 4xy^3 + 2) \\
&= 6x^2y^5 - 2xy^3 - 8 + 7x^2y^5 + 4xy^3 - 2 \\
&= 6x^2y^5 + 7x^2y^5 - 2xy^3 + 4xy^3 - 8 - 2 \\
&= 13x^2y^5 + 2xy^3 - 10
\end{aligned}$$

✎ *Pencil Problem #9* ✎

9a. Subtract:

$(17x^3 - 5x^2 + 4x - 3) - (5x^3 - 9x^2 - 8x + 11)$

9b. Subtract $-5a^2b^4 - 8ab^2 - ab$ from $3a^2b^4 - 5ab^2 + 7ab$.

Objective 2.R.10: Multiply polynomials when neither is a monomial.

✔ *Solved Problem #10*

10. Multiply: $(3x+2)(2x^2-2x+1)$

$$(3x+2)(2x^2-2x+1)$$
$$=3x(2x^2-2x+1)+2(2x^2-2x+1)$$
$$=6x^3-6x^2+3x+4x^2-4x+2$$
$$=6x^3-2x^2-x+2$$

Pencil Problem #10

10. Multiply: $(x-3)(x^2+2x+5)$

Objective 2.R.11: Find the domain of a rational function.

✔ *Solved Problem #11*

11. Find the domain of f if $f(x)=\dfrac{x-5}{2x^2+5x-3}$.

The domain is all real numbers except those which make the denominator zero. Set the denominator equal to zero and solve.

$$2x^2+5x-3=0$$
$$(2x-1)(x+3)=0$$

$2x-1=0$ or $x+3=0$

$x=\frac{1}{2}$ $\quad$ $x=-3$

Domain of $f=(-\infty,-3)\cup\left(-3,\frac{1}{2}\right)\cup\left(\frac{1}{2},\infty\right)$.

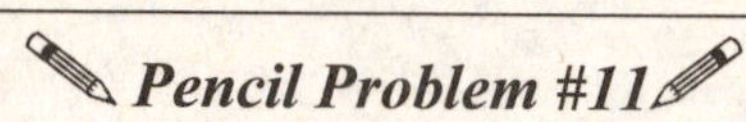

Pencil Problem #11

11. Find the domain of f if $f(x)=\dfrac{2x}{(x+5)^2}$.

Objective 2.R.12: Simplify complex rational expressions by multiplying by 1.

✔ Solved Problem #12

12a. Simplify: $\dfrac{\dfrac{x}{y}-1}{\dfrac{x^2}{y^2}-1}$

Multiply the numerator and denominator by the LCD of y^2.

$$\frac{\dfrac{x}{y}-1}{\dfrac{x^2}{y^2}-1}=\frac{y^2}{y^2}\cdot\frac{\dfrac{x}{y}-1}{\dfrac{x^2}{y^2}-1}=\frac{\dfrac{x\cdot y^2}{y}-1\cdot y^2}{\dfrac{x^2\cdot y^2}{y^2}-1\cdot y^2}$$

$$=\frac{xy-y^2}{x^2-y^2}$$

$$=\frac{y(x-y)}{(x+y)(x-y)}$$

$$=\frac{y}{x+y}$$

Pencil Problem #12

12a. Simplify: $\dfrac{4+\dfrac{2}{x}}{1-\dfrac{3}{x}}$

12b. Simplify: $\dfrac{\dfrac{1}{x+7}-\dfrac{1}{x}}{7}$

Multiply the numerator and denominator by the LCD of $x(x+7)$.

$$\frac{\frac{1}{x+7}-\frac{1}{x}}{7}=\frac{x(x+7)}{x(x+7)}\cdot\frac{\frac{1}{x+7}-\frac{1}{x}}{7}$$

$$=\frac{\frac{x(x+7)}{x+7}-\frac{x(x+7)}{x}}{7x(x+7)}$$

$$=\frac{x-(x+7)}{7x(x+7)}$$

$$=\frac{x-x-7}{7x(x+7)}$$

$$=\frac{-7}{7x(x+7)}$$

$$=\frac{-1}{x(x+7)}$$

$$=-\frac{1}{x(x+7)}$$

12b. Simplify: $\dfrac{\dfrac{1}{x-2}}{1-\dfrac{1}{x-2}}$

Objective 2.R.13: Find the domain of square root functions.

✔ ***Solved Problem #13***

13. Find the domain of $f(x)=\sqrt{9x-27}$.

The radicand, $9x-27$, must be nonnegative.

$9x-27\geq 0$

$9x\geq 27$

$x\geq 3$

The domain of f is $[3,\infty)$.

Pencil Problem #13

13. Find the domain of $f(x)=\sqrt{3x+15}$.

Objective 2.R.14: Complete the square of a binomial.

✔ *Solved Problem #14*

14. What term should be added to the binomial $x^2 + 10x$ so that it becomes a perfect square trinomial? Write and factor the trinomial.

The coefficient of the x-term of $x^2 + 10x$ is 10.

Half of 10 is 5, and 5^2 is 25, which should be added to the binomial.

The result is a perfect square trinomial.

$x^2 + 10x + 25 = (x+5)^2$

Pencil Problem #14

14. What term should be added to the binomial $x^2 - 14x$ so that it becomes a perfect square trinomial? Write and factor the trinomial.

Answers for Pencil Problems:

1. 31

2a.

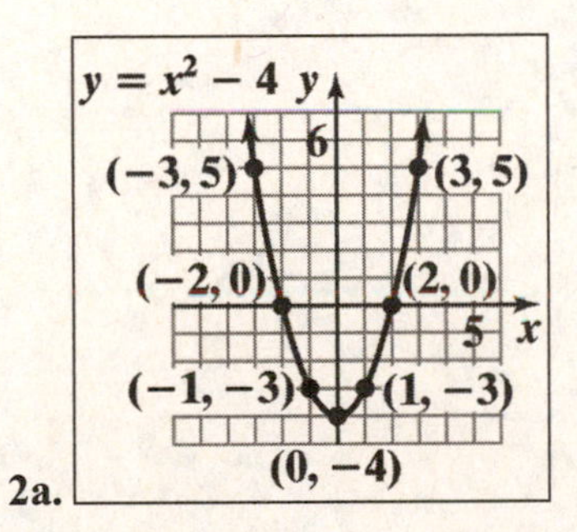

2b.

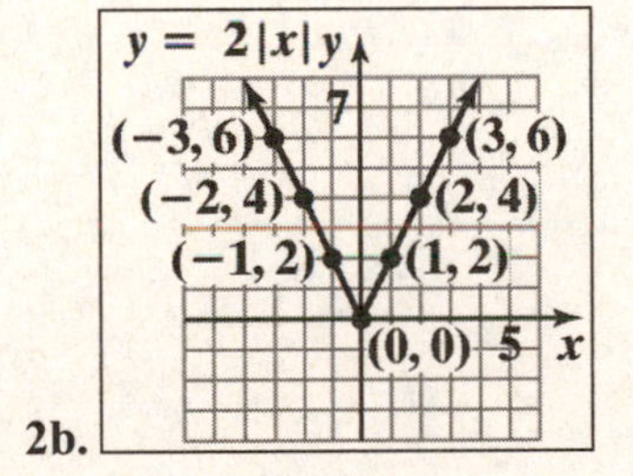

3a. $p = \dfrac{T-D}{m}$

3b. $I = \dfrac{E}{R+r}$

4a. $4x^2 + 4xy + y^2$

4b. $y^2 - 10y + 25$

5a. $\dfrac{1}{x-3}$

5b. $\dfrac{4}{x}$

6a.

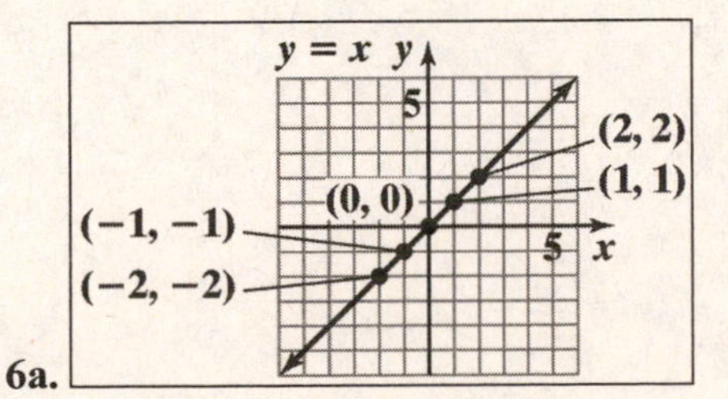

6b.

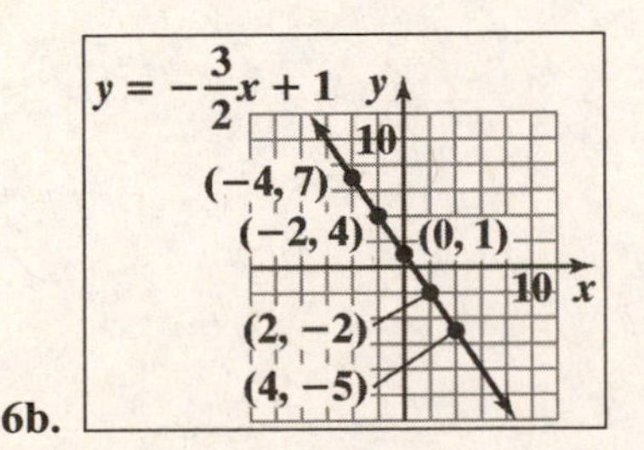

7a. $p = \dfrac{T - D}{m}$ **7b.** $I = \dfrac{E}{R + r}$

8a. $11x^3 + 7x^2 - 12x - 4$ **8b.** $9x^2y - 6xy$

9a. $12x^3 + 4x^2 + 12x - 14$ **9b.** $8a^2b^4 + 3ab^2 + 8ab$

10. $x^3 - x^2 - x - 15$

11. $(-\infty, -5) \cup (-5, \infty)$

12a. $\dfrac{4x + 2}{x - 3}$ **12b.** $\dfrac{1}{x - 3}$

13. $[-5, \infty)$

14. Add 49; $x^2 - 14x + 49 = (x - 7)^2$

Chapter 3
Polynomial and Rational Functions

Guided Practice:

☐ Review each of the following ***Solved Problems*** and complete each ***Pencil Problem***.

Objective 3.R.1: Use the vocabulary of polynomial functions.

✔ *Solved Problem #1*

1. Determine the coefficient of each term, the degree of each term, the degree of the polynomial, the leading term, and the leading coefficient of the polynomial $8x^4y^5 - 7x^3y^2 - x^2y - 5x + 11$.

The degree of the polynomial is 9, the leading term is $8x^4y^5$, and the leading coefficient is 8.

Term	Coefficient	Degree
$8x^4y^5$	8	$4+5=9$
$-7x^3y^2$	−7	$3+2=5$
$-x^2y$	−1	$2+1=3$
$-5x$	−5	1
11	11	0

Pencil Problem #1

1. Determine the coefficient of each term, the degree of each term, the degree of the polynomial, the leading term, and the leading coefficient of the polynomial $x^3y^2 - 5x^2y^7 + 6y^2 - 3$.

Objective 3.R.2: Evaluate polynomial functions.

✔ *Solved Problem #2*

2. Find $f(2)$ for the polynomial function $f(x) = 4x^3 - 3x^2 - 5x + 6$.

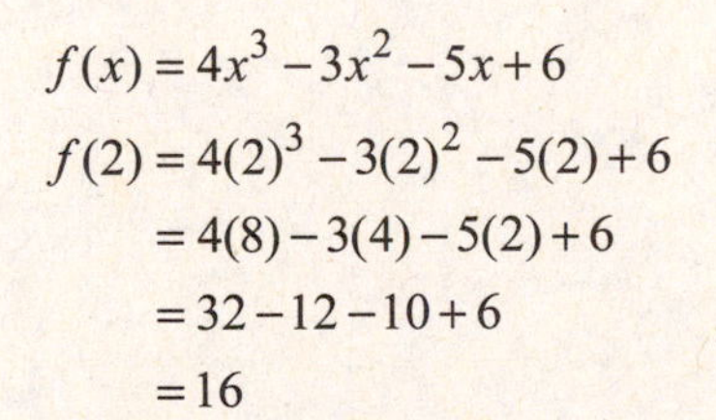

$$f(x) = 4x^3 - 3x^2 - 5x + 6$$
$$f(2) = 4(2)^3 - 3(2)^2 - 5(2) + 6$$
$$= 4(8) - 3(4) - 5(2) + 6$$
$$= 32 - 12 - 10 + 6$$
$$= 16$$

Pencil Problem #2

2. Find $g(3)$ for the polynomial function $g(x) = 2x^3 - x^2 + 4x - 1$.

Objective 3.R.3: Use a general factoring strategy.

✔ *Solved Problem #1*

3a. Factor: $3x^3 - 30x^2 + 75x$

$$\begin{aligned} 3x^3 - 30x^2 + 75x &= 3x(x^2 - 10x + 25) \\ &= 3x(x-5)^2 \end{aligned}$$

3b. Factor: $3x^2y - 12xy - 36y$

$$\begin{aligned} 3x^2y - 12xy - 36y &= 3y(x^2 - 4x - 12) \\ &= 3y(x+2)(x-6) \end{aligned}$$

3c. Factor: $16a^2x - 25y - 25x + 16a^2y$

$$\begin{aligned} 16a^2x - 25y - 25x + 16a^2y &= 16a^2x + 16a^2y - 25y - 25x \\ &= (16a^2x + 16a^2y) + (-25y - 25x) \\ &= 16a^2(x+y) - 25(y+x) \\ &= 16a^2(x+y) - 25(x+y) \\ &= (x+y)(16a^2 - 25) \\ &= (x+y)(4a+5)(4a-5) \end{aligned}$$

Pencil Problem #1

3a. Factor: $x^3 - 16x$

3b. Factor: $4x^2 + 25y^2$

3c. Factor: $x^4 - xy^3 + x^3y - y^4$

3d. Factor: $x^2 - 36a^2 + 20x + 100$

$$\begin{aligned} x^2 - 36a^2 + 20x + 100 &= x^2 + 20x + 100 - 36a^2 \\ &= (x^2 + 20x + 100) - 36a^2 \\ &= (x+10)^2 - 36a^2 \\ &= (x + 10 + 6a)(x + 10 - 6a) \end{aligned}$$

3d. Factor: $x^2 - 12x + 36 - 49y^2$

Answers for Pencil Problems:

1. The coefficient of x^3y^2 is 1 and the degree is 5. The coefficient of $-5x^2y^7$ is –5 and the degree is 9. The coefficient of $6y^2$ is 6 and the degree is 2. The coefficient of –3 is –3 and the degree is 0. The degree of the polynomial is 9. The leading term is $-5x^2y^7$ and the leading coefficient is –5.

2. 56

3a. $x(x+4)(x-4)$

3b. Prime

3c. $(x+y)(x-y)(x^2+xy+y^2)$

3d. $(x-6+7y)(x-6-7y)$

Chapter 4
Exponential and Logarithmic Functions

Guided Practice:

☐ Review each of the following ***Solved Problems*** and complete each ***Pencil Problem***.

Objective 4.R.1: Evaluate exponential expressions.

✔ *Solved Problem #1*

1a. Evaluate: 6^2

$$6^2 = 6 \cdot 6$$
$$= 36$$

1b. Evaluate: $(-1)^4$

$$(-1)^4 = (-1)(-1)(-1)(-1)$$
$$= 1$$

1c. Evaluate: -1^4

$$-1^4 = -(1 \cdot 1 \cdot 1 \cdot 1)$$
$$= -1$$

Pencil Problem #1

1a. Evaluate: 9^2

1b. Evaluate: $(-4)^3$

1c. Evaluate: $(-5)^4$

Objective 4.R.2: Use the zero-exponent rule.

✔ *Solved Problem #2*

2. Use the zero-exponent rule to simplify: -5^0

$$-5^0 = -(5^0)$$
$$= -1$$

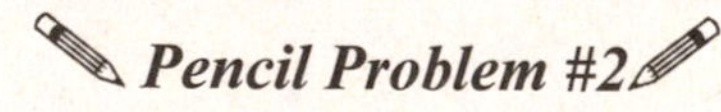

Pencil Problem #2

2. Use the zero-exponent rule to simplify: $(-4)^0$

Objective 4.R.3: Use the negative-exponent rule.

✔ *Solved Problem #3*

3a. Use the negative-exponent rule to write 5^{-2} with a positive exponent. Simplify, if possible.

$$5^{-2} = \frac{1}{5^2} = \frac{1}{25}$$

Pencil Problem #3

3a. Use the negative-exponent rule to write -5^{-2} with a positive exponent. Simplify, if possible.

✔ *Solved Problem #3*

3b. Use the negative-exponent rule to write $\frac{1}{5x^{-2}}$ with a positive exponent. Simplify, if possible.

$$\frac{1}{5x^{-2}} = \frac{x^2}{5}$$

Pencil Problem #3

3b. Use the negative-exponent rule to write $\frac{1}{5^{-3}}$ with a positive exponent. Simplify, if possible.

Objective 4.R.4: Use the product rule.

✔ *Solved Problem #4*

4a. Multiply using the product rule: $b^6 \cdot b^5$

$$b^6 \cdot b^5 = b^{6+5} = b^{11}$$

Pencil Problem #4

4a. Multiply using the product rule: $3x^4 \cdot 2x^2$

✔ *Solved Problem #4*

4b. Multiply using the product rule: $\left(4x^3y^4\right)\left(10x^2y^6\right)$

$$\left(4x^3y^4\right)\left(10x^2y^6\right) = 4 \cdot 10 \cdot x^3 \cdot x^2 \cdot y^4 \cdot y^6 = 40x^{3+2}y^{4+6} = 40x^5y^{10}$$

Pencil Problem #4

4b. Multiply using the product rule: $\left(-2y^{10}\right)\left(-10y^2\right)$

Objective 4.R.5: Use the quotient rule.

✔ ***Solved Problem #5***

5. Divide using the quotient rule: $\dfrac{27x^{14}y^8}{3x^3y^5}$

$$\frac{27x^{14}y^8}{3x^3y^5} = \frac{27}{3}x^{14-3}y^{8-5}$$
$$= 9x^{11}y^3$$

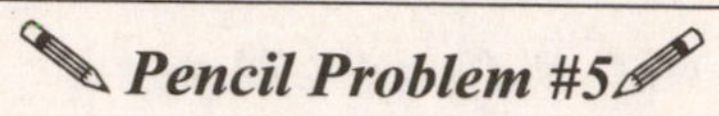

5. Divide using the quotient rule: $\dfrac{50x^2y^7}{5xy^4}$

Objective 4.R.6: Use the power rule.

✔ ***Solved Problem #6***

6. Simplify $(b^{-3})^{-4}$ using the power rule.

$$(b^{-3})^{-4} = b^{(-3)(-4)}$$
$$= b^{12}$$

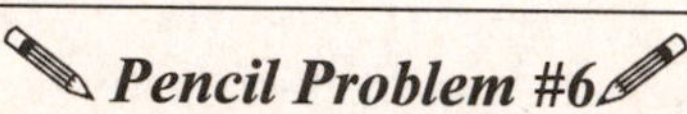

6. Simplify $\left(7^{-4}\right)^{-5}$ using the power rule.

Objective 4.R.7: Model data with linear functions and make predictions.

✔ *Solved Problem #7*

7. The data for the life expectancy for American women are displayed as a set of six points in the scatter plot. Also shown is a line that passes through or near the six points.

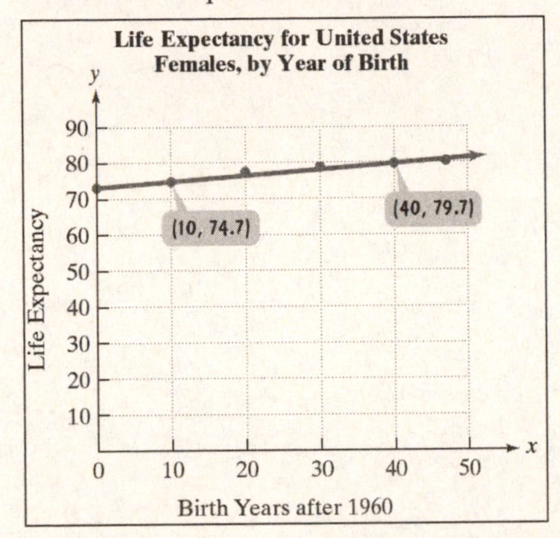

Use the data points labeled by the voice balloons to write the slope-intercept form of the equation of this line. Round the slope to two decimal places. Then use the linear function to predict the life expectancy of an American woman born in 2020.

First, find the slope.

$$m = \frac{79.7 - 74.7}{40 - 10} = \frac{5}{30} \approx 0.17$$

Then use the slope and one of the points to write the equation in point-slope form.

Using the point (10, 74.7):

$$y - y_1 = m(x - x_1)$$
$$y - 74.7 = 0.17(x - 10)$$
$$y = 0.17x + 73$$
$$f(x) = 0.17x + 73$$

Next, since 2020 is 60 years after 1960, substitute 60 into the function: $f(60) = 0.17(60) + 73 = 83.2$.

This means that the life expectancy of American women in 2020 is predicted to be 83.2 years.

Answers vary due to rounding and choice of point. If point (40, 79.7) is chosen, $f(x) = 0.17x + 72.9$ and the life expectancy of American women in 2020 is predicted to be 83.1 years.

✎ *Pencil Problem #7* ✎

7. In 2005 there were 40.8 million smartphones sold in the United States. In 2010 there were 296.6 million smartphones sold. A scatterplot of smartphone sales from 2004 to 2010 is shown, where x represents the number of years after 2004 and y represents the number of smartphones sold, in millions.

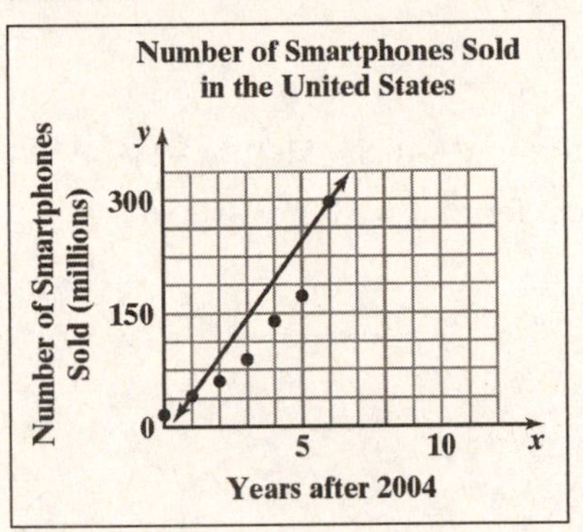

A line is drawn through the two points that show the number of smartphones sold in 2005 and 2010. Use the coordinates of these points to write the line's equation in point-slope form and slope-intercept form.

Answers for Pencil Problems:

1a. 81 **1b.** −64 **1c.** 625

2. 1

3a. $-\frac{1}{25}$ **3b.** 125

4a. $6x^6$ **4b.** $20y^{12}$

5. $10xy^3$

6. 7^{20}

7. Point-Slope form: $y-40.8=51.16(x-1)$ or $y-296.6=51.16(x-6)$,
Slope-Intercept form: $y=51.16x-10.36$

Chapter 5
Systems of Equations and Inequalities

Guided Practice:

☐ Review each of the following ***Solved Problems*** and complete each ***Pencil Problem***.

Objective 5.R.1: Determine whether an ordered pair is a solution of an equation.

✔ *Solved Problem #1*

1a. Determine whether the ordered pair $(3,-2)$ is a solution of the equation $x-3y=9$.

$$x-3y=9$$
$$3-3(-2)=9$$
$$3+6=9$$
$$9=9,\ \text{true}$$

$(3,-2)$ is a solution.

1b. Determine whether the ordered pair $(-2,3)$ is a solution of the equation $x-3y=9$.

$$x-3y=9$$
$$-2-3(3)=9$$
$$-2-9=9$$
$$-11=9,\ \text{false}$$

$(-2,3)$ is not a solution.

Pencil Problem #1

1a. Determine whether the ordered pair $(0,6)$ is a solution of the equation $y=2x+6$.

1b. Determine whether the ordered pair $(2,-2)$ is a solution of the equation $y=2x+6$.

Objective 5.R.2: Use set-builder notation.

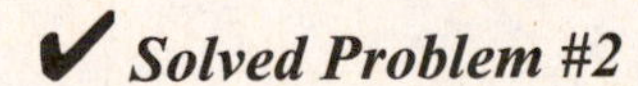

Solved Problem #2

2. Use the roster method to list the elements in the set.
$\{x \mid x \text{ is a natural number less than } 6\}$

{1, 2, 3, 4, 5}

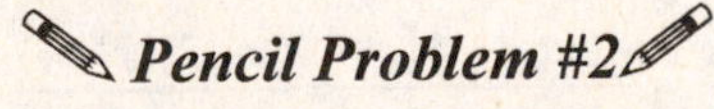

Pencil Problem #2

2. Use the roster method to list the elements in the set.
$\{x \mid x \text{ is an integer between } -8 \text{ and } -3\}$

Objective 5.R.3: Add and subtract rational expressions with different denominators.

Solved Problem #3

3a. Add: $\frac{7}{6x^2}+\frac{2}{9x}$

The LCD is $18x^2$.

Rewrite each rational expression with the LCD as the denominator.

$$\frac{7}{6x^2}+\frac{2}{9x}=\frac{7}{6x^2}\cdot\frac{3}{3}+\frac{2}{9x}\cdot\frac{2x}{2x}$$
$$=\frac{21}{18x^2}+\frac{4x}{18x^2}$$
$$=\frac{4x+21}{18x^2}$$

3b. Add: $\frac{x}{x-4}+\frac{x-2}{x+4}$

The LCD is $(x-4)(x+4)$.

Rewrite each rational expression with the LCD as the denominator.

$$\frac{x}{x-4}+\frac{x-2}{x+4}=\frac{x(x+4)}{(x-4)(x+4)}+\frac{(x-2)(x-4)}{(x+4)(x-4)}$$
$$=\frac{x^2+4x}{(x-4)(x+4)}+\frac{x^2-6x+8}{(x-4)(x+4)}$$
$$=\frac{2x^2-2x+8}{(x-4)(x+4)}$$
$$=\frac{2(x^2-x+4)}{(x-4)(x+4)}$$

Pencil Problem #3

3a. Add: $\frac{3}{5x^2}+\frac{10}{x}$

3b. Add: $\frac{4}{x-2}+\frac{3}{x+1}$

3c. Subtract: $\dfrac{2x-3}{x^2-5x+6}-\dfrac{x+4}{x^2-2x-3}$

The LCD is $(x-3)(x+1)(x-2)$.

Rewrite each rational expression with the LCD as the denominator.

$$\frac{2x-3}{x^2-5x+6}-\frac{x+4}{x^2-2x-3}$$
$$=\frac{2x-3}{(x-3)(x-2)}-\frac{x+4}{(x-3)(x+1)}$$
$$=\frac{(2x-3)(x+1)}{(x-3)(x-2)(x+1)}-\frac{(x+4)(x-2)}{(x-3)(x+1)(x-2)}$$
$$=\frac{2x^2-x-3}{(x-3)(x-2)(x+1)}-\frac{x^2+2x-8}{(x-3)(x+1)(x-2)}$$
$$=\frac{2x^2-x-3-(x^2+2x-8)}{(x-3)(x+1)(x-2)}$$
$$=\frac{2x^2-x-3-x^2-2x+8}{(x-3)(x+1)(x-2)}$$
$$=\frac{x^2-3x+5}{(x-3)(x+1)(x-2)}$$

3c. Subtract: $\dfrac{3y+7}{y^2-5y+6}-\dfrac{3}{y-3}$

3d. Perform the indicated operations:

$$\frac{y-1}{y-2}+\frac{y-6}{y^2-4}-\frac{y+1}{y+2}$$

The LCD is $(y+2)(y-2)$.

Rewrite each rational expression with the LCD as the denominator.

$$\frac{y-1}{y-2}+\frac{y-6}{y^2-4}-\frac{y+1}{y+2}$$

$$=\frac{y-1}{y-2}+\frac{y-6}{(y-2)(y+2)}-\frac{y+1}{y+2}$$

$$=\frac{(y-1)(y+2)}{(y-2)(y+2)}+\frac{y-6}{(y+2)(y-2)}-\frac{(y+1)(y-2)}{(y+2)(y-2)}$$

$$=\frac{y^2+y-2}{(y-2)(y+2)}+\frac{y-6}{(y+2)(y-2)}-\frac{y^2-y-2}{(y+2)(y-2)}$$

$$=\frac{y^2+y-2+y-6-(y^2-y-2)}{(y+2)(y-2)}$$

$$=\frac{y^2+y-2+y-6-y^2+y+2}{(y+2)(y-2)}$$

$$=\frac{3y-6}{(y+2)(y-2)}$$

$$=\frac{3(y-2)}{(y+2)(y-2)}$$

$$=\frac{3}{y+2}$$

3d. Perform the indicated operations:

$$\frac{3}{5x+6}-\frac{4}{x-2}+\frac{x^2-x}{5x^2-4x-12}$$

Objective 5.R.4: Use long division to divide by a polynomial containing more than one term.

✔ Solved Problem #4

4a. Simplify: $\dfrac{\dfrac{x+1}{x-1}-\dfrac{x-1}{x+1}}{\dfrac{x-1}{x+1}+\dfrac{x+1}{x-1}}$

The LCD of the numerator is $(x+1)(x-1)$.

The LCD of the denominator is $(x+1)(x-1)$.

$$\dfrac{\dfrac{x+1}{x-1}-\dfrac{x-1}{x+1}}{\dfrac{x-1}{x+1}+\dfrac{x+1}{x-1}}=\dfrac{\dfrac{(x+1)(x+1)}{(x-1)(x+1)}-\dfrac{(x-1)(x-1)}{(x+1)(x-1)}}{\dfrac{(x-1)(x-1)}{(x+1)(x-1)}+\dfrac{(x+1)(x+1)}{(x-1)(x+1)}}$$

$$=\dfrac{\dfrac{x^2+2x+1}{(x-1)(x+1)}-\dfrac{x^2-2x+1}{(x+1)(x-1)}}{\dfrac{x^2-2x+1}{(x+1)(x-1)}+\dfrac{x^2+2x+1}{(x-1)(x+1)}}$$

$$=\dfrac{\dfrac{x^2+2x+1-(x^2-2x+1)}{(x+1)(x-1)}}{\dfrac{x^2-2x+1+x^2+2x+1}{(x+1)(x-1)}}$$

$$=\dfrac{\dfrac{x^2+2x+1-x^2+2x-1}{(x+1)(x-1)}}{\dfrac{x^2-2x+1+x^2+2x+1}{(x+1)(x-1)}}$$

$$=\dfrac{\dfrac{4x}{(x+1)(x-1)}}{\dfrac{2x^2+2}{(x+1)(x-1)}}$$

$$=\dfrac{4x}{(x+1)(x-1)}\cdot\dfrac{(x+1)(x-1)}{2x^2+2}$$

$$=\dfrac{4x}{2(x^2+1)}$$

$$=\dfrac{2x}{x^2+1}$$

Pencil Problem #4

4a. Simplify: $\dfrac{\dfrac{x+2}{x-2}-\dfrac{x-2}{x+2}}{\dfrac{x-2}{x+2}+\dfrac{x+2}{x-2}}$

4b. Simplify: $\dfrac{1-4x^{-2}}{1-7x^{-1}+10x^{-2}}$

Rewrite the expression without negative exponents.

Then multiply the numerator and denominator by the LCD of x^2.

$$\frac{1-4x^{-2}}{1-7x^{-1}+10x^{-2}}=\frac{1-\frac{4}{x^2}}{1-\frac{7}{x}+\frac{10}{x^2}}$$

$$=\frac{x^2}{x^2}\cdot\frac{1-\frac{4}{x^2}}{1-\frac{7}{x}+\frac{10}{x^2}}$$

$$=\frac{1\cdot x^2-\frac{4\cdot x^2}{x^2}}{1\cdot x^2-\frac{7\cdot x^2}{x}+\frac{10\cdot x^2}{x^2}}$$

$$=\frac{x^2-4}{x^2-7x+10}$$

$$=\frac{(x+2)(x-2)}{(x-5)(x-2)}$$

$$=\frac{x+2}{x-5}$$

4b. $\dfrac{3a^{-1}+3b^{-1}}{4a^{-2}-9b^{-2}}$

Answers for Pencil Problems:

1a. $(0,6)$ is a solution **1b.** $(2,-2)$ is not a solution

2. $\{-7,-6,-5,-4\}$

3a. $\dfrac{3+50x}{5x^2}$ **3b.** $\dfrac{7x-2}{(x-2)(x+1)}$ **3c.** $\dfrac{13}{(y-3)(y-2)}$ **3d.** $\dfrac{x^2-18x-30}{(5x+6)(x-2)}$

4a. $\dfrac{4x}{x^2+4}$ **4b.** $\dfrac{3ab(b+a)}{(2b+3a)(2b-3a)}$

Chapter 6
Matrices and Determinants

Objective 6.R.1: Use commutative, associative, and distributive properties.

✔ *Solved Problem #1*

1a. Use an associative property to write an equivalent expression and simplify: $6+(12+x)$

$$6+(12+x)=(6+12)+x$$
$$=18+x$$

1b. Use the distributive property to write an equivalent expression: $-4(7x+2)$

$$-4(7x+2)=-28x-8$$

Pencil Problem #1

1a. Use an associative property to write an equivalent expression and simplify: $-7(3x)$

1b. Use the distributive property to write an equivalent expression: $-(3x-6)$

Objective 6.R.2: Find multiplicative inverses.

✔ *Solved Problem #2*

2a. Find the multiplicative inverse of 7.

The multiplicative inverse of 7 is $\frac{1}{7}$ because $7\cdot\frac{1}{7}=1$.

$\frac{1}{7}$

2b. Find the multiplicative inverse of $-\frac{7}{13}$.

The multiplicative inverse of $-\frac{7}{13}$ is $-\frac{13}{7}$

because $\left(-\frac{7}{13}\right)\left(-\frac{13}{7}\right)=1$.

$-\frac{13}{7}$

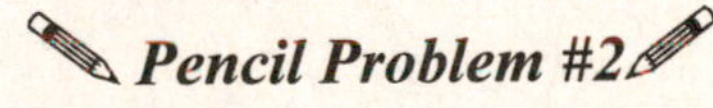

Pencil Problem #2

2a. Find the multiplicative inverse of -10.

2b. Find the multiplicative inverse of $\frac{1}{5}$.

Answers for Pencil Problems:

1a. $-21x$ **1b.** $-3x+6$

2a. $-\frac{1}{10}$ **2b.** 5

Chapter 8
Sequences, Induction, and Probability

Guided Practice:

☐ Review each of the following ***Solved Problems*** and complete each ***Pencil Problem***.

Objective 8.R.1: Find the power of a product.

✔ *Solved Problem #1*

1. Simplify the expression using the products-to-powers rule: $\left(-4x^5y^{-1}\right)^{-2}$

$$\left(-4x^5y^{-1}\right)^{-2} = (-4)^{-2}\left(x^5\right)^{-2}\left(y^{-1}\right)^{-2}$$
$$= \frac{1}{(-4)^2} \cdot x^{-10} \cdot y^2$$
$$= \frac{y^2}{16x^{10}}$$

Pencil Problem #1

1. Simplify the expression using the products-to-powers rule: $\left(5x^3y^{-4}\right)^{-2}$

Objective 8.R.2: Simplify exponential expressions.

✔ *Solved Problem #2*

2a. Simplify: $\left(-3x^{-6}y\right)\left(-2x^3y^4\right)^2$

$$(-3x^{-6}y)(-2x^3y^4)^2 = (-3x^{-6}y)(-2)^2(x^3)^2(y^4)^2$$
$$= -3 \cdot x^{-6} \cdot y \cdot 4 \cdot x^6 \cdot y^8$$
$$= -12 \cdot x^{-6+6} \cdot y^{1+8}$$
$$= -12x^0y^9$$
$$= -12y^9$$

Pencil Problem #2

2a. Simplify: $\left(2a^5\right)\left(-3a^{-7}\right)$

2b. Simplify: $\left(\dfrac{10x^3y^5}{5x^6y^{-2}}\right)^2$

$$\left(\frac{10x^3y^5}{5x^6y^{-2}}\right)^2 = \left(2x^{3-6}y^{5+2}\right)^2$$
$$= \left(2x^{-3}y^7\right)^2$$
$$= 4x^{-6}y^{14}$$
$$= \frac{4y^{14}}{x^6}$$

2b. Simplify: $\left(\dfrac{-15a^4b^2}{5a^{10}b^{-3}}\right)^3$

Objective 8.R.3: Reduce or simplify fractions.

✔ *Solved Problem #3*

3a. Reduce $\dfrac{10}{15}$ to its lowest terms.

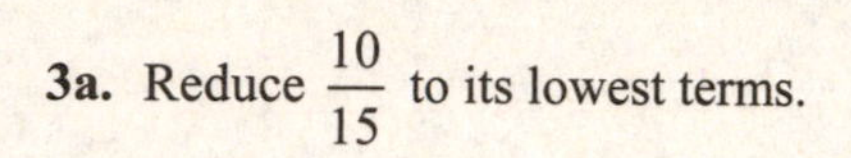

$$\frac{10}{15} = \frac{2 \cdot \cancel{5}}{3 \cdot \cancel{5}} = \frac{2}{3}$$

3b. Reduce $\dfrac{13}{15}$ to its lowest terms.

13 and 15 share no common factors (other than 1).

Therefore, $\dfrac{13}{15}$ is already reduced to its lowest terms.

Pencil Problem #3

3a. Reduce $\dfrac{35}{50}$ to its lowest terms.

3b. Reduce $\dfrac{120}{86}$ to its lowest terms.

Objective 8.R.4: Multiply fractions.

✔ *Solved Problem #4*

4a. Multiply $\frac{4}{11} \cdot \frac{2}{3}$.

If possible, reduce the product to its lowest terms.

$$\frac{4}{11} \cdot \frac{2}{3} = \frac{4 \cdot 2}{11 \cdot 3}$$
$$= \frac{8}{33}$$

4b. Multiply $\left(3\frac{2}{5}\right)\left(1\frac{1}{2}\right)$.

If possible, reduce the product to its lowest terms.

$$\left(3\frac{2}{5}\right)\left(1\frac{1}{2}\right) = \frac{17}{5} \cdot \frac{3}{2}$$
$$= \frac{51}{10}$$
$$= 5\frac{1}{10}$$

Pencil Problem #4

4a. Multiply $\frac{3}{8} \cdot \frac{7}{11}$.

If possible, reduce the product to its lowest terms.

4b. Multiply $\left(3\frac{3}{4}\right)\left(1\frac{3}{5}\right)$.

If possible, reduce the product to its lowest terms.

Objective 8.R.5: Add and subtract fractions with identical denominators.

✔ *Solved Problem #5*

5a. Perform the indicated operation: $\frac{2}{11} + \frac{3}{11}$

$$\frac{2}{11} + \frac{3}{11} = \frac{2+3}{11}$$
$$= \frac{5}{11}$$

5b. Perform the indicated operation: $\frac{5}{6} - \frac{1}{6}$

$$\frac{5}{6} - \frac{1}{6} = \frac{4}{6}$$
$$= \frac{2}{3}$$

Pencil Problem #5

5a. Perform the indicated operation: $\frac{7}{12} + \frac{1}{12}$

5b. Perform the indicated operation: $\frac{11}{18} - \frac{4}{18}$

Objective 8.R.6: Add and subtract fractions with unlike denominators.

Solved Problem #6

6a. Perform the indicated operation: $\frac{1}{2}+\frac{3}{5}$

$$\begin{aligned}\frac{1}{2}+\frac{3}{5}&=\frac{1\cdot 5}{2\cdot 5}+\frac{3\cdot 2}{5\cdot 2}\\&=\frac{5}{10}+\frac{6}{10}\\&=\frac{5+6}{10}\\&=\frac{11}{10}\end{aligned}$$

6b. Perform the indicated operation: $3\frac{1}{6}-1\frac{11}{12}$

$$\begin{aligned}3\frac{1}{6}-1\frac{11}{12}&=\frac{19}{6}-\frac{23}{12}\\&=\frac{19\cdot 2}{6\cdot 2}-\frac{23}{12}\\&=\frac{38}{12}-\frac{23}{12}\\&=\frac{15}{12}\\&=\frac{5}{4}\\&=1\frac{1}{4}\end{aligned}$$

Pencil Problem #6

6a. Perform the indicated operation: $\frac{3}{8}+\frac{5}{12}$

6b. Perform the indicated operation: $3\frac{3}{4}-2\frac{1}{3}$

Answers for Pencil Problems:

1. $\frac{y^8}{25x^6}$

2a. $-\frac{6}{a^2}$ **2b.** $-\frac{27b^{15}}{a^{18}}$

3a. $\frac{7}{10}$ **3b.** $\frac{60}{43}$

4a. $\frac{21}{88}$ **4b.** 6

5a. $\frac{2}{3}$ **5b.** $\frac{7}{18}$

6a. $\frac{19}{24}$ **6b.** $\frac{17}{12}$ or $1\frac{5}{12}$